Technology & Management

(a multidisciplinary collection)

Editors: Shahryar Sorooshian

Amin Teyfouri

Siti Aissah bt Mad Ali

Table of Contents

Micro Robots and Automated Accessories 7

Review on Warehouse Administration 16

Advanced Manufacturing Technology and Process 27

Indicators and Models of Sustainable Development 50

Review on Supply Chain Management: 4th Party Logistic 62

Improving HSE in Construction Area by Using Virtual Reality (VR), a Comprehensive Review 78

A Review on Micro-Manufacturing, Deforming Processes and Their Prominent Issues 92

Robotics Application In Civil and Management Engineering 117

Automated Robotic Maintenance for Industry 128

Managing the Chain of Suppliers 145

Proposal of Study on Determinants of Consumer's Purchasing Intention towards Green Products 160

Leadership and Characteristics of Women Principals In Secondary Schools In Johor Bahru 171

A Comparison between Green Concrete and Conventional Concrete 185

Sustainable Development of Manufacturing 207

Overview of Medical Waste Management in Malaysia 221

From Idea to Enterprise: The Business Concept Statement 234

Business Feasibility Assessment on Energy Storage Capacitor: A case of Commercializing University Research 242

Elements of Total Quality Management 250

Micro Robots and Automated Accessories

Amin Teyfouri[1a], Parisa Keyhani Boroojeni [2b]

[1] *Engineering Faculty, UPM, Malaysia*

[2] *Engineering Faculty, UPM, Malaysia*

Academic Editor: *Shahryar Sorooshian*

ABSTRACT: In this point of time, the demand for micro components and products has been notably expanded. Hence the importance of micro-robots for making these products is a growing matter in the contemporary world. This paper takes account of the Micro-robots and its principle applications, including micro-assembly, micro-handling and sensor robots.

Keywords: Actuators; Micro-assembly; Micro-handling; Micro-robots and Sensors

1.0 INTRODUCTION

As a matter of fact, staying in the competitive market needs some factors. Implementations of advanced manufacturing technologies and as sequence of it use the right robot for achieving a competitive price and high quality material is the vital matter in firms [1]. A robot can be defined based on its capability to respond according to the environment and the degree of intelligence [2]. Macrobiotics is the field of miniature robotics, especially mobile robots with characteristic sizes less than 1 mm. The term can also be used for robots capable of handling micrometer size components .Section 2 reviews the classification of robots. Section

3 depicts the Micro-assembly. Section 4 provides an illustration of micro handling and its principles. Section 5 describes the several kinds of sensors in micro-robots. Section 6 Suggests the then Section 7 summarizes this paper.

2.0 ROBOTS CLASSIFICATION

Generally speaking, a robot is a programmable machine that is supposed to automatically carry out a diversity of actions [2]. For the most part, for achieving a successful implementation, Robotics application should be clarify in advanced while it would be along with spending much more time on programming robots besides, capability to merely do the certain operations owing to preprogramming. Thus the flexible robots are the suitable cases for active manufacturing as they have ability to adjustment to the new operations [3]. Robots are categorized to three groups according to their size as Macro robots, Micro robots and Nano robots.

In this point of time, Micro robots come to the fore in micro manufacturing. They are far smaller robots with the ability to conduct micro scale thing. In some point of view, micro robots are tiny centimeter size robots that are made by an amalgam of micro electrons and conventional methods, while the other groups are of the opinion that micro robots are sub millimeter or millimeter robots which are built by the technology of micro-machines [4]. By the way of illustration, the new micro sheet forming machine which its set-up is flexible so the power can be change while machine set-ups is unchanged. Moreover, it's a cheap, stable and bench top machine [5]. Designing a hot embossing machine of polymeric micro tube that is adaptable to different features and shapes (FIG 1.) [6]. Currently, micro-electromechanical systems (MEMS) are defined as the technology of small parts that have a widespread use in variety of fields such as medical, communication, military, car components, aeronautics, multimedia, etc [7] [8]. Nevertheless there exist several barriers for implementation of them in many fields since they are including micro-component that should be assembled in several stages that need flexible micro handling systems plus experts by relevant skills [8].

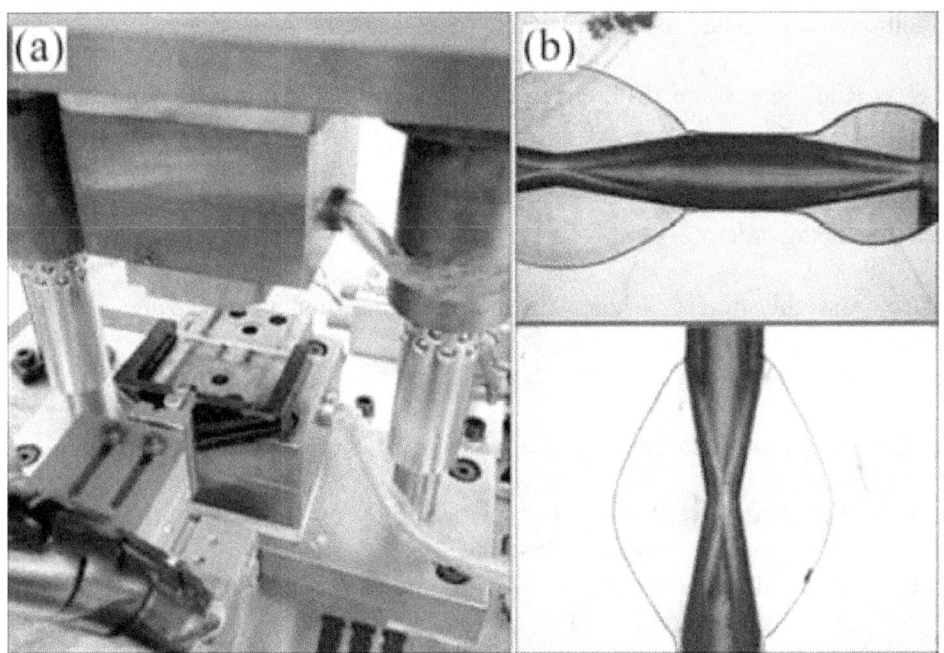

Figure 1: Micro-tube hot-embossing machine and its sample of formed parts [6]

Nano robotics is defines as a technology for making robots in nanometers scale which is 0.1 to 10 millimeters [9]. Nano robots are still under development, by acquiring the sophisticated technologies; scientist could create a nano robotics devises in different field such as industrial and medical by complement the miniature device which moves through body.

3.0   MICRO-ASSEMBLY

Micro-assembly is the assembly of several micro-components that are each smaller than 1 millimeters and usually are seen around 100 micrometers [7]. In other words, micro robotics is combination of techniques and theory such as sensor fusion, motion control, etc. by MEM technology for improving human's life [10].There are a notable progress in micro-assembly such as gripping, positioning, bonding, and handling parts. Micro-assembly has some same features with the Traditional one, including task planning, grasping, and part orientation and so on [11]. Nonetheless, there are some differences between the macro-assembly and the micro-assembly such as the different requirements for accuracy and manipulation. Micro robots are conducted for doing micro-assembly which can gain the highest efficiency in comparison with other assembly tools. Whereas it needs rather much more time and expertize. In the micro-assembly system, a workforce or, preferably an inspection system is conducted for observing the components during the operation which means access state-of-the-art technologies [11]. For assembly of micro-parts it is highly essential to find viable sequences since each of them illustrate the possibility of implementation of the process planning [12].

4.0 MICRO-HANDLING

Micro-handling is the other application of micro-robots. In fact, it is a manipulation of small components. As micro components are delicate they should be preserve by frequent manipulation and tidiness, thus a neat workplace is compulsory [8]. Picking up the micro parts and releasing them is major process of micro-handling, so find the well position for set the devices with short distance for moving and traveling to gain great accuracy is a key issue. Today, automated micro-handling technology comes to the fore due to the much level of accuracy. On the other hand it brings some drawbacks such as the limitation of accessing and having accurate and flexible micro-handling machines and insufficient standards that can lead to wasting time and also resources by relevant workforces. In micro-handling magnetic, friction and also pneumatic plays an undeniable and vital role [8].

5.0 SENSORS IN MICRO-ROBOTS

As a matter of fact an indispensable role of sensors in robotics is taken for granted. Sensors have task for collecting info about the environment. These tasks include temperature measurement, avoiding collisions, distance measurement, obstacle avoidance, etc. Undoubtedly, sensors are so important in micro-world application since they have to be accurate and precise and it is noted that Most of the micro-robots have a number of sensors due to their measurement. Basically, sensors use an electronic circuit and a transducer; that is, transducer is a tool that converts form of energy to electrical signal such as photocurrent, thermistor, electrochemical. Combination of numerous signals can make a model that is sensor fusion. Sensors normally categorized to two groups: external and internal or active and passive. These sensors are most common for using in robotics.

Infrared sensors

Infrared sensors are an electronic device for detecting an aspect of its surroundings, their environment and / or detect infrared radiation. Infrared sensors measure the heat of an object, it also recognizes the motion. Many of these types of sensors to measure the infrared radiation only, instead of the transmission, and then known as passive (PIR) sensors infrared. This radiation is invisible to the eye but can be detected by an infrared sensor that accepts and interprets it [13].

Tactile sensors

Tactile sensor usually mentions to a transducer that is sensitive to pressure, touch, or force [reference Tactile Sensing—From Humans to Humanoids]. Touch and tactile sensors are devices for measuring some parameters between an object and a sensors. Tactile sensors can be used to detect a wide range of stimulus for the presence or absence of an object from a contact picture. A tactile sensor comprises a touch sensitive set of sites, the sites for measuring more than one property.

Resistive sensor

A resistive sensor has ability for converting mechanical change to electrical signal; that is, a resistive sensor is a type common instrumentation in electromechanical device or transducer. Resistance changes due to material or geometry changes. The potentiometer is simplest resistive sensor. Resistive sensors usually are combination of Wheatstone bridges.

Orientation sensor

The orientation sensors are combined gyroscope and accelerometers for using data in real time. Due to these specification this sensor, this sensor become perfect transducer for a functional electrical stimulation (FES) [14] and accuracy of orientation in medical imaging [15]. However, it has major drawbacks because of expensive price.

Inertial sensors

Inertial sensors is kind of device for measure and detect vibration, acceleration, shock, rotation, etc. Inertial sensors stem from MEMS inertial sensors for using some devices such as smart phones because of high cost performance [16] [17], the dimension of these sensors are different from a few square mm to 50 cm. Noise normally sets the limit to resolution of an inertial sensor and it is impossible that resolve signal.

6.0   ACTUATORS IN MICRO-ROBOTS

Actuators are normally mechanical device for producing a movement in a system, process or mechanism; in other words, actuators are the suppliers of the signals to the robots. Obviously, micro-actuators are necessary for micro-robots for creating their movement such as electrostatic actuators, shape-memory-alloy composite micro-actuators, piezoelectric actuators, micro-pneumatic actuators, micro-motors [18]. Piezoelectric actuators have a little amount, small quantity of heat and high speed response. The most important of these actuators are explained; micro-motors and piezoelectric actuators

1. Micro-motors: There are two basic types of micro-motors: piezoelectric and electromagnetic. Piezoelectric micro-motors make the change in shape of a piezoelectric material. The signal is produced from the ultrasonic vibrations of the material when applying an electrical field.

2. Piezoelectric Actuators: These actuators generate signal in the sub-nanometer range. The actuators use piezoelectric materials and the frequencies resulted from solid-state crystalline properties. Piezoelectric Actuators typically have a long life time without any maintenance and short response time.

7.0 CONCLUSION

This paper demonstrates the several principals in micro-robots and its accessories. As the growth in utilizing the micro products, firms should allocate much more money and time to find the suitable technology and device for implementation to being competitiveness. Hence work in micro-robot is a must and requires extra efforts. There are a glimpse of micro robots its assembly and material handling in addition to its sensors and actuators. Future studies should be more concentrates on the robots to find and categorized the devices for executing the right robots in different firm by various products.

REFERENCES

[1] F. K. H. H. G. B. L Alting, "Micro Engineering," vol. 52, no. 2, 2003.

[2] J. O. MATTHEW J. LIBERATORE, "Planning a Successful Robotics Implementation," *JOURNAL OF OPERATIONS MANAGEMENT,* vol. 5, pp. 247-257, 1985.

[3] L. W. R. G. Mohammed, "Integrated image processing and path planning for robotic sketching," *8th CIRP Conference on Intelligent Computation in Manufacturing Engineering,* vol. 12, pp. 199-204, 2013.

[4] O. I. a. I. L. Edwin W. H. Jager, "Microrobots for Micrometer-Size Objects in Aqueous Media: Potential Tools for Single-Cell Manipulation," *American Association for the Advancement of Science,* vol. 288, no. 5475, pp. 2335-2338, 2000.

[5] Y. Q. Akhtar Razul Razali, "A Review on Micro-manufacturing, Micro-forming and their Key Issues," *Procedia Engineering,* vol. 53, pp. 665-672, 2013.

[6] A. L. K. Y. Y. Q. A. R. Z.-c. F. Jie ZHAO, "Machine and tool development for forming of polymeric tubular micro-components," *Trans. Nonferrous Met. Soc. China,* vol. 22, pp. s214-s221, 2012.

[7] M. e. G. r. P. nique Gendreau n, "Modular architectureofthemicrofactoriesforautomaticmicro-assembl," *Robotics andComputer-IntegratedManufacturing,* vol. 26, pp. 354-360, 2010.

[8] R. L.-T. ,. G.-D. ,. C. R.-V. A.J. Sanchez-Salmeron, "Recent development in micro-handling systems for micro-manufacturing," *Journal of Materials Processing Technology,* vol. 167, pp. 499-507, 2005.

[9] R. C. Santosh Kumar Verma, "Nanorobotics in dentistry e A review," *Indian Journal of Dentistry,* pp. 1-9, 2013.

[10] S. M. S. M.Sunitha Reddy, "Riview on Microrobot," 2011.

[11] D. O. P. a. H. E. Stephanou, "Micro and Mesoscale Robotic Assembly," *Journal of Manufacturing ProcessES,* vol. 6, no. 1, pp. 52-71, 2004.

[12] C. Son, "Intelligent control planning strategies with neural network/fuzzy coordinator and sensor fusion for robotic part macro/micro-assembly tasks in a partially unknown environment," *International Journal of Machine Tools & Manufacture,* vol. 44, pp. 1667-1681, 2004.

[13] J. W. J Tegin, "Tactile sensing in intelligent robotic manipulation," 2005.

[14] S. P. G. M. D. R. W. S. J. W. M. Scott Simcox, "Performance of orientation sensors for use with a functional electrical stimulation mobility system," vol. 38, 2005.

[15] G. M. T. A. H. G. a. R. W. P. Richard James Housden, "Calibration of an orientation sensor for freehand 3D ultrasound and its use in a hybrid acquisition system," 2008.

[16] n. Y. ,. H. ,. Y. Thanh TrungNgo, "The largestinertialsensor-basedgaitdatabaseandperformance evaluationofgait-basedpersonalauthentication," vol. 47, 2014.

[17] K. N. R. T.Kobayashi, "Rotation in variant feature extraction from 3-D acceleration signals," 2011.

[18] X. X. S. Li Sui, "Piezoelectric A ctuator Design and Application on Active Vibration Control," 2012.

*****

Review on Warehouse Administration

Milad Ahmadi

UPM University, Serdang, Malaysia

Academic Editor: *Amin Teyfouri*

ABSTRACT: This paper explains how to provide an efficient and economic buffer stock of materials so that users are provided with level of service commensurate with the maximum objectives. These explanation also to ensure proper administration of the warehouse to control handling, storage, preservation and distribution / delivery of purchased materials to user/s. The scope also includes the applicable to the warehouse personnel who handle all types of purchased materials during receiving, storage and distribution of these materials to users from the warehouse.

Keywords: Storage of Material; Issuance of Material; Expediting of Purchase Order; Branch Office; Document Administration; Disposal of Scrape Material

1.0 INTRODUCTION

In many industries and supply chains, inventory is one of the dominant costs. In the United States, for example, over a trillion dollars is invested in inventory. For many managers, effective supply chain management is synonymous with reducing inventory levels in the supply chain [1]. Of course, this is very simplistic view of supply chain management. In fact, the goal of effective inventory management in the supply chain is to have the correct inventory at the right place at the right time to minimize system costs while satisfying customer service requirements. Unfortunately, managing inventory in complex supply chain is typically difficult, and inventory-related decisions can have the significant impact on the customer service level and supply chain system wide cost [2].

The warehouse environment has drastically changed. As a result of customer demand for faster and more accurate deliveries, the emphasis is now focused on accelerating product flow and managing costs. To complicate the issue, customer-mandated product customization, such as order assembly, packaging requirements, compliance labeling, and other delivery requirements, are becoming the norm. Your challenge is to manage this increasingly complex use of your facilities and the greater demands on your employees' time efficiently and effectively [3].

1.0 STORAGE OF MATERIAL

In terms of storage of material, upon verification of purchased materials, all conforming purchased materials shall be kept at the warehouse to allow for easy retrieval. Then, ***Bin Card** shall be used to identify stock items in the warehouse, as well as to monitor the flow of the stock items, i.e. incoming to and outgoing from the warehouse. While in the other side, the person-in-charge (PIC) warehouse and the Store Clerk shall ensure that the stocks are stored in the proper condition to prevent damage or deterioration. The PIC of warehouse or Store Clerk also shall be responsible to assess the condition of stocks every three (3) months, which includes quality and quantity verification of stocks against Bin Card. At the same time, they must be responsible to designate and identify area to store non-conforming purchased products or products that has not been inspected or verified.

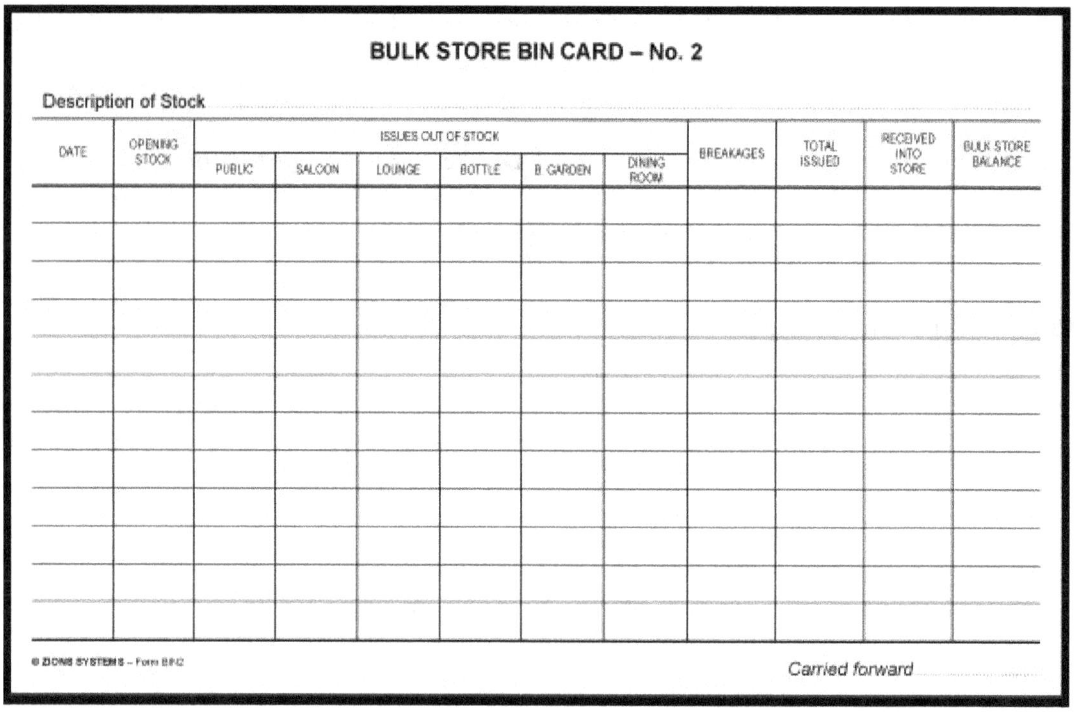

BULK STORE BIN CARD – No. 2

Description of Stock ....

DATE	OPENING STOCK	ISSUES OUT OF STOCK						BREAKAGES	TOTAL ISSUED	RECEIVED INTO STORE	BULK STORE BALANCE
		PUBLIC	SALOON	LOUNGE	BOTTLE	B. GARDEN	DINING ROOM				

© ZIONS SYSTEMS – Form BIN2

Carried forward ....

Figure 1: Example of Bin Card

2.0 ISSUANCE OF MATERIAL

Upon receipt of approved Material Request Form (MRF) from the requestor, the PIC, Warehouse or his Store Clerk shall check the availability of the material requested at the warehouse [4].

2.1 Material Request Form (MRF)

A *material request form* lists the items to be picked from inventory and used in the production process or in the provision of a service to a customer, usually for a specific job. The form usually has three purposes [5][6]:

- To pick items from stock
- To relieve the inventory records in the amount of the items picked
- To charge the targeted job for the cost of the items requisitioned

The form can also be used as the basis for the reordering of any inventory items that are not currently in stock.

The information most commonly found on a material requisition form includes:

- Header section   : Job number to be charged
- Header section   : Date of requisition
- Header section   : Date by which inventory is required
- Main body   : Item number or description to be pulled from stock
- Main body   : Unit quantity to be pulled from stock
- Footer section   : Authorization signature line

If the materials are to be delivered to a specific location, there may also be space in the header in which to identify the delivery location.

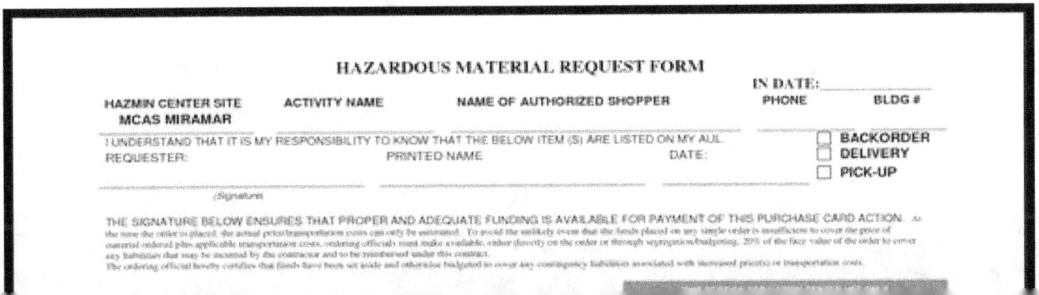

Figure 2.1: Example of Material Request Form

Then, if the material is available, the PIC, Warehouse or the assigned Store Clerk shall issue the material together with Warehouse Dispatch Note approved by the PIC, Warehouse or any of Store Clerk. If the material is not available, the PIC, Warehouse or their assigned Store Clerk shall determine whether the material requested is a stock item by referring to the *J.D Edwards Warehouse Management System. But if the material is a stock item, the PIC, Warehouse or his assigned Store Clerk shall raise purchase requisition to their Head of Department (HOD) for approval.

2.1 J.D EDWARDS WAREHOUSE MANAGEMENT SYSTEM

J.D Edwards Warehouse Management System provides the flexible, automated support you need for better customer service and lower operating costs. Advanced rules-based stock location and directed movement logic helps you dramatically improve warehouse labor efficiency and space usage. And, full integration with sales, purchasing, manufacturing, and transportation enables you to meet your customers' preferences for how they want their orders sourced, made, packaged, and shipped.

3.0 EXPEDITING OF PURCHASE ORDER

The person-in-charge (PIC) of warehouse shall assign a Store Clerk to be responsible for expediting of Purchase Order (PO). The assigned Store Clerk also shall monitor the delivery date of material/s as stated in the PO end of each month. In other side, a reminder letter or e-mail shall be sent to the supplier stating the PO number and date required if the supplier fail to deliver within the required date.

3.1 Purchase Order (PO)

A **purchase order (PO)** is a commercial document issued by a buyer to a seller, indicating types, quantities, and agreed prices for products or services the seller will provide to the buyer. Sending a PO to a supplier constitutes a legal offer to buy products or services. Acceptance of a PO by a seller usually forms a one-off contract between the buyer and seller, so no contract exists until the purchase order is accepted.

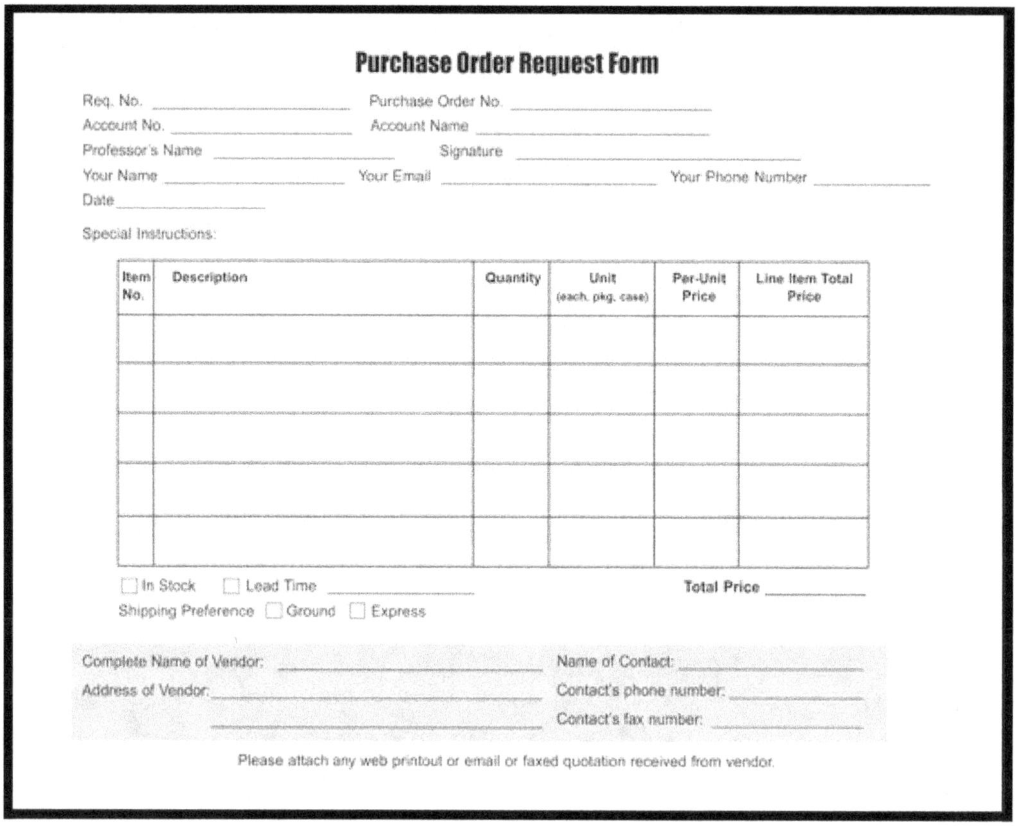

Purchase Order Request Form

Req. No. ____ Purchase Order No. ____
Account No. ____ Account Name ____
Professor's Name ____ Signature ____
Your Name ____ Your Email ____ Your Phone Number ____
Date ____

Special Instructions:

Item No.	Description	Quantity	Unit (each, pkg, case)	Per-Unit Price	Line Item Total Price

☐ In Stock ☐ Lead Time ____ **Total Price** ____
Shipping Preference ☐ Ground ☐ Express

Complete Name of Vendor: ____ Name of Contact: ____
Address of Vendor: ____ Contact's phone number: ____
____ Contact's fax number: ____

Please attach any web printout or email or faxed quotation received from vendor.

Figure 3.1: Example of Purchase Order Form (PO)[7]

4.0 BRANCH OFFICE WAREHOUSE ADMINISTRATION

The central warehouse shall receive all materials for distribution to the branch office warehouses. The materials shall be transported to the respective warehouse via lorry or any transportation that available and a Dispatch Note shall be issued for every issuance to the branch warehouses. Then, the branch store clerk shall check the Dispatch Note and the material, before endorsing the Dispatch Note. The original Dispatch Note shall be filed and the copy shall be submitted to the central warehouse. In the event that the material has to be delivered directly to the regional office warehouse by the suppliers, the branch store clerk shall check the Delivery Order (DO) and the material before endorsing the DO with the

"GOOD RECEIVED IN GOOD ORDER" stamp. The original DO shall be submitted to the central warehouse and the photocopy to be filed at the respective warehouse. Upon receipt of approved Material Request Form (MRF) or System Generated MRF, the Store Clerk shall check the stock and issue the material with the Dispatch Note. Besides, Bin card shall be maintained for stock item monitoring and branch warehouse staff s h all update the material that have been issued in the JDE System.

5.0   DOCUMENT ADMINISTRATION

The processes of Document Administration need the person-in-charge (PIC) of warehouse or the assigned Store Clerk shall submit the following documents via transmittal note to the Accounting & Finance Department upon completion of the receiving of material activity for payment process [8]:

• Original Delivery Order (DO)

• Original Invoice

• Copy of General Journal Batch Report (where applicable)

The transmittal note for payment shall be checked by person-in-charge (PIC), Warehouse and verified by the PIC, Procurement prior approval by Head of Department (HOD), Procurement & Contract before submitting to Accounting & Finance. The following documents shall be kept by the PIC, Warehouse or his assigned Store Clerk:

•Copy of Delivery Order (DO)

•Copy of Purchase Order (PO)

• Relevant certificate from the supplier (where applicable) finally, the PIC, Warehouse or his assigned Store Clerk shall maintain the Bin Card

6.0 DISPOSAL OF SURPLUS MATERIAL

For return of surplus materials to central warehouse by the contractors, it will be kept at the designated warehouse store yard. In a while, person-in-charge of warehouse will arrange for a Third Party inspection (if necessary) to determine the status of surplus material either in useable condition or scrap which prior recommendation for disposal [9]. A Capital Expenditure Disposal Form shall be prepared by the person-in-charge of the warehouse and will be reviewed and recommended for disposal by the highest authority i.e. Head of Department (HOD), Procurement & Contracts, prior approval by Chief Financial Officer (CFO), Senior General Manager, Operations & Maintenance (SGMOM), Chief Operating Officer (COO) and Managing Director (MD). Upon approval, the disposal of these scrap materials will be processed and handled by the Procurement Section of Procurement & Contracts Department for tendering and award purposes [10].

7.0 CONCLUSION

As the conclusion, no matter how simple or complex the administration is, the goal of a warehouse management system remains the same – which is to provide management with the information it needs to efficiently control the movement of materials within a warehouse.

Warehouse management is not limited to documenting the delivery of raw materials and the movement of those materials into operational process. The movement of those materials as they go through the various stages of the operation is also important. Typically known as a goods or work in progress warehouse, tracking materials as they are used to create finished goods also helps to identify the need to adjust ordering amounts before the raw materials inventory gets dangerously low or is inflated to an unfavorable level.

Finally, warehouse management has to do with keeping accurate records of finished goods that are ready for shipment. This often means posting the production of newly completed goods to the warehouse totals as well as subtracting the most recent shipments of finished goods to buyers. When the company has a return policy in place, there is usually a sub-category contained in the finished goods warehouse to account for any returned goods that are reclassified as refurbished or second grade quality. Accurately maintaining figures on the finished goods warehouse makes it possible to quickly convey information to sales personnel as to what is available and ready for shipment at any given time.

REFERENCES

1. Roland Holten : Specification of Management Views In Information Warehouse Project, Volume 28,Issue 7, October 2003
2. J.P.van den Berg, W.H.M. Zijm : Models For Warehouse Management: Classification And Example, International Journal of Production Economics, Volume 59, Issues 1–3, March 1999, Pages 519-528
3. Klaus Moeller: Increasing Warehouse Order Picking Performance Sequence Optimization, Procedia - Social and Behavioral Sciences, Volume 20, 2011, Pages 177-185
4. K.W Chau, Ying Cao, M Anson, Jianping Zhang : Application of Data Warehouse and Decision Support System In Construction Management, Automation in Construction, Volume 12, Issue 2, March 2003, Pages 213-224

5. Hamid R. Nemati, David M. Steiger, Lakshmi S. Iyer, Richard T. Herschel : Knowledge Warehouse: An Architectural Integration of Knowledge Management, Decision Support, Artificial Intelligence And Data Warehousing , Decision Support Systems, Volume 33, Issue 2, June 2002, Pages 143-161

6. Yeu-Shiang Huang, Do Duy, Chih-Chiang Fang : Efficient Maintenance of Basic Statistical Function In data Warehouse, Decision Support Systems, In Press, Corrected Proof, 2013

7. Sae-yeon Roh, Hyun-mi Jang, Chul-hwan Han: Warehouse Location Decision Factors in Humanitarian relief Logistics, The Asian Journal of Shipping and Logistics, Volume 29, Issue 1, April 2013, Pages 103-120

8. Géraldine Strack, Yves Pochet: An Integrated Model For Warehouse And Inventory Planning , European Journal of Operational Research, Volume 204, Issue 1, 1 July 2010, Pages 35-50

9. Jinxiang Gu, Marc Goetschalckx, Leon F. McGinnis: Research On Warehouse Operation: A Comprehensive Review, European Journal of Operational Research, Volume 177, Issue 1, 16 February 2007, Pages 1-21

10. Peter Baker, Marco Canessa: Warehouse Design : A Structured Approach, European Journal of Operational *Research, Volume 193, Issue 2, 1 March 2009, Pages 425-436*

*****

Advanced Manufacturing Technology and Process

Morteza tolou[a], Shahrzad Abbasi[b]

UPM University, Serdang, Malaysia

Academic Editor: *Amin Teyfouri*

ABSTRACT: Modern manufacturing is at the frontier of new technologies, products and ways of working. Our future lies in a mixed and balanced economy with manufacturing and services reinforcing each other. Manufacturing creates wealth, sustains jobs and is central to national economic success and also has been the foundation of a nation's strength as a trading nation in both the past and the present. The propensities in manufacturing in examine from few but advanced industrialized nations like Japan, US, Germany, Britain etc. with more emphasis on Japan manufacturing. This paper describes the historical background and types of manufacturing industries, and the current trends in manufacturing such as technological innovation, market–oriented manufacturing, and Global manufacturing. We also deduced the in-depth of the current trend in manufacturing such as product and process innovation, diversification and multiple product (small-batch production), and multinational enterprises. We highlighted the challenges faced by manufacturing industries such as large underutilized investment, people and skills, restructuring, poor manufacturing performance and global market environment. Based on the challenges highlighted, we strongly recommended that long-term planning, skill development; right market framework and good knowledge of new methods should be implemented to

create enormous advantage to the global economy and also enhances the livelihood of the people globally.

Keywords: Current trend in manufacturing; Technological innovation; Market oriented manufacturing; Global manufacturing

1.0 INTRODUCTION

No matter what industry you work in, it is important to keep up with the current trends. These trends cover the employment rate, turnover rate, economy and numerous other factors. In fact if you are unaware of the industry trends in manufacturing your business might collapse because you are ignorant of your competitor's strengths.

Being alert of the industrial propensities helps to keep you competitive with other businesses in your industry and also provides advantages.

Due to the rapid movement in industrial trends in various industries it seems like a huge task trying to keep up with the game and such noticeable trends are customers' expectation, products, technology, etc. the tragic of all is that trends that are affecting other industries can directly affect your line of business, so it is best to be acquainted with most of the current manufacturing trends.

Current trends of globalization, mass-customization and increased competition are leading industries toward networked organizations, such as virtual organizations and dynamic business ecosystems. Increasingly, products must be rapidly adapted to customer needs, leading to faster innovation cycles and more complex concurrent engineering. In many industry sectors, a demand for increased value creation on the supplier and sub-suppliers side of the networks is observed, emphasizing knowledge content and services. This requires significant growth in inter- and intra-enterprise knowledge sharing, management services, and in collaborative work environments and work management services.

2.0 HISTORICAL BACKGROUND

The manufacturing industry is, historically, regarded as one of the key drivers in any regional or national economy. This stems from the fact that it was the development of the advanced manufacturing sector from the 18th century which propelled the world into the modern era and which provided a financial and technological basis for improved standards of living, modern capitalism and ultimately advanced international transactions [3]. The experience of the United States and Western Europe which achieved high levels of socio-economic development, consequent upon their industrial revolutions in the 18th and 19th centuries was for many years regarded as a 'model' for less developed countries. Emulation has however proven to be difficult owing to changing social and economic circumstances at the local, national and global levels and differing resource, financial and capacity endowments. So prior to the 1970s the world's economy was characterized by a sense of permanence and relatively fixed and predictable actions [4]. Industrial power was concentrated in the developed Western nation and countries in the developing world tended to play the role of being both markets and suppliers of raw materials. In the 1970s, as a result of a series of global crises, namely the devaluation of the dollar, a petrol price crisis and the collapse of much of the older, less efficient industries in the core economies, significant changes started taking place, namely:

- Deindustrialization in the older industrial cores, particular those areas which relied on heavy industry, such as iron and steel and textiles

- The partial relocation of the locus of activities for heavy and labor-intensive countries to countries in the developing world and the so called Newly Industrializing countries in particular

- The continued exercise of control from the core cities / countries in which the global economic head-quarters are based

- The movement of the world system to one marked by instability and unpredictability as investors ceaselessly invest and disinvest in their search for the most profitable locations

- The focusing of the 'new high tech.' industries on locations within the developed world

- The increased importance of global trade and the 'weakening' of global borders as a result of actions of the World Trade Organization, cheaper bulk transport and world-wide subscription to the principle of globalization.

- There is an associated belief that the geography of fixed places of production has been replaced with geography of flow and interconnection in a fluid world environment in which little is now fixed or permanent,

- Industrial activity is increasing focusing on defined industrial districts and clusters, characterized by concentrations of like complementary and competing and supporting industries e.g. vehicle clusters and linked supply and support firms [7].

One of the key hall-marks of the preceding has been the dominance of the situation of fluidity and the lack of permanence. Notions that manufacturing firms are locality bound in terms of their location, raw materials, labor and market have now been replaced by a situation in which markets are global, and locations decisions and actions are not bound to isolated spatial areas. While this does mean that consumers now have access to better quality and priced products, producers and workers now face increased competition from firms around the world and investors think far broader than their home area [9]. As a direct result, industrial economies are volatile and subject to rapid change.

3.0 TYPES OF MANUFACTURING INDUSTRIES

3.1 Industrial Classification

Industries are usually classified into three categories (A.G.B. Fisher): (I) *primary industry* (production relying on nature) — agriculture, forestry, and fishing and mining (which is often included in the next category); (II) *secondary industry* (production of tangible goods (products) through conversion of raw materials) — manufacturing, construction, and electricity, gas and water (which are often included in the next category); and (III) *tertiary industry* (intangible service production) — transportation and communication, wholesale and retail trade, finance, insurance and real estate, services, and government (and no classified establishments). As the economies of a nation have grown, the industrial pattern shifts from I to II, and further from II to III (Petty and Clark's law). A country in which the percentage of labor population in the tertiary industry has exceeded 50 is recognized as an economically advanced country that has entered into the post-industrial or information society [8].

3.2 Types of Manufacturing Systems

There are nine various manufacturing system groups, they are as follows and include many various manufacturing systems.

- Manufacturing strategy and paradigms.
- Manufacturing systems design and operations.
- Sustainable manufacturing.
- Quality management.
- Automation, control systems, human-machine interaction.

- Product development.

- Supply chain management and logistics.

- Manufacturing information systems.

- Micro and Nano manufacturing systems

Manufacturing strategy and paradigms are flexible, reconfigurable and changeable manufacturing systems. These manufacturing systems include rapid manufacturing, lean manufacturing, and virtual enterprises. Rapid manufacturing is used for large products that have layer-based manufacturing from metals, plastics, or composite material. Lean manufacturing is when the manufacturing is removing the waste, usually time, that happens during manufacturing. Virtual enterprises manufacturing system is the ability to virtually join or combine efforts with other enterprises to improve manufacturing [7].

Manufacturing systems design and operations are process planning, production planning and controls, modeling, simulation, and virtual manufacturing. These manufacturing systems will actually be helpful when started to manufacture or a change is needed in manufacturing. Planning and modeling can be done with any manufacturing type. Virtual manufacturing is the use of audiovisual features to simulate as actual manufacturing environment, then using the virtual manufacturing to see the results.

Sustainable manufacturing includes the life cycle of products and systems, and designs for environments and sustainability. Sustainable manufacturing systems have the goal to improve the use of resources and make the product more environment and health friendly and well and make the product more marketable.

Quality management includes the manufacturing systems that are known to improve the quality of the product as the main goal. They are manufacturing systems like product and process quality, quality function deployment, and quality by design and six sigma.

Automation, control systems, and human-machine interaction are manufacturing systems that have either robots or machines doing the bulk of the work. The various manufacturing systems here are agent-based systems, distributed and integrated control systems, intelligent systems, emergent systems, reconfigurable control, robotics, collaborative robots and human-machine interactions.

Product development manufacturing systems are systems that focus on the product development in manufacturing. There are product families, reverse engineering concept development, product design and integration with manufacturing systems and product life cycle [12].

Supply chain management and logistics manufacturing systems are systems that will get the best in the supplies that are needed to manufacture a product. Some supply chain manufacturing systems are global supply chain, dynamic supply chains, modeling and optimization.

Manufacturing information systems allow business to gain the information needed to improve the manufacturing of the products. Various manufacturing systems here include the Internet, web-based systems, enterprise resource planning, automatic data capture and enterprise modeling.

Micro and Nano- manufacturing systems are systems that are related to micro and nano-developing products. Many are technological products that are small products to manufacture. There are, systems issues related to micro fluidics, nano electronics, nano systems, micro electromechanical systems, nano materials, interconnects, and energy, chemical and biological devices [15].

All of these various systems will provide a business with the ability to improve the manufacturing of the many diverse products being manufactured in the world. Most companies will use more than one of the manufacturing systems in their manufacturing.

There are large manufacturing companies that will combine a series of the various manufacturing systems. Then once the manufacturing systems have accomplished the goal that has been made, a new manufacturing system may be used to make improvements elsewhere in the company [13].

4.0 CURRENT TREND IN MANUFACTURING INDUSTRIES

Manufacturing is very important in creating the wealth of a nation. In other to observed, understand and describe the current trend in manufacturing, few industrialized or leading countries in manufacturing are put into considerations such as the Japan, US, UK and Germany, with more emphasis on Japan manufacturing.

Manufacturing industries have developed very rapidly in Japan and that has been the key to the creation of her national wealth. Japanese industries have advanced manufacturing technology for thorough rationalization of manufacturing processes. The trend in manufacturing is been described based on technological innovation, market–oriented manufacturing, and International / Global manufacturing.

4.1 Technological Innovation

This term was first used by the Austrian economist J.A Schumpeter of which he identified five kinds of innovation, but for the sake of our technical/ term paper, we consider two-related and most significant which are 'product innovation' and 'process innovation'.

4.1.1 Product Innovation

It is the market introduction of a product (goods or services) that is new or significantly changed in terms of core characteristics, technical specifications, built-in software or any other immaterial component or the intended use or ease of use [5].

Industrial companies innovate primarily by introducing new products on the market. Electricals, Home equipment, and an automobile manufacturing industries pioneered innovation. Innovating is a way of scoring points against the competition.

Considering Japanese manufacturing industries, in recent years many original Japanese products have been developed, mass produced, and have gained a dominant share of the world market. Examples of these products are digital watches, videotape recorders, and others including home entertainment computers and software.

Since there is much severe competition in Japan, Japanese companies finds it essential to continuously modify, adapt, and refine products so as to meet the competition. This is true for all products, those invented and developed in other countries and those indigenously created in Japan. The diagram below shows the radically new products launched on the market from 1998 to 2000 generated nearly 9% of the manufacturing industry turnover.

The innovation concept is defined in the OECD Oslo Manual [3]. The concept makes international comparisons possible. The OECD definition is more restrictive than the definition qualifying for the public funding. The innovations may also be the result of improvements or of the supply of new services for the products (Service' innovations).

Improving the product quality and opening up new markets rank at the top of the main objectives of innovation activities. In-house R&D turns out to be the main sources of information assisting innovation activities. 51.2% of the firms that are engaged in innovation carry out joint R&D with consultancy firms, and 52.3% of the firms with which Turkish firms co-operate are in the EU countries.

In the majority of the manufacturing sectors, more than 50% of the total sales are derived from technologically new and improved products. Only 19% of the firms have had patent applications with a return of very few patented inventions [3]. A correlation analysis of basic indicators of innovation activities shows that, for instance, sales of new products, R&D expenditures, and firm sizes correlate only weakly.

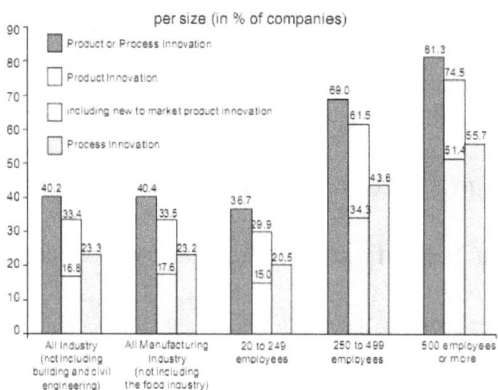

Figure 1.Innovation according to the international definition of the Oslo Manual

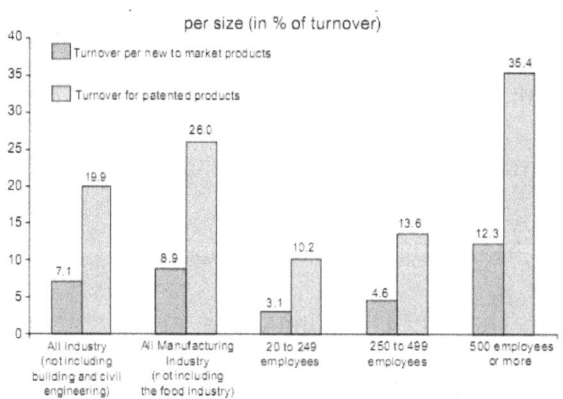

Figure 2.The impact of innovation on corporate business

4.1.2 Process Innovation

This is also known as Factory automation. It comprises the introduction of a new or significantly changed manufacturing process, service supply method or product delivery method into the company [8]. The result must have a significant impact on the level of production, the product quality or the production and distribution costs.

In other to cope with the shortage of skillful workers, and to overcome the high wage rate, the present goal of the modern manufacturing plant in Japan is the 'unmanned factory'. Comparing wage rates in 1988 (with 100 used as the standard), the relative wage rates in the following countries were: Japan 100; US 100; Germany 125; United Kingdom 78; and Korea 30. The unmanned factory is considered as the only way for Japan to compete with Newly Industrializing Economies (NIES) such as Korea, Taiwan, Hong Kong, Singapore, etc., which still have low labor costs [12].

Since in 1990, the average direct labor cost as a percentage of the total cost was 11.7% for small/medium sized manufacturing firms, unmanned factories can save this amount in wages. Machine-tool makers now produce several flexible manufacturing systems (FMS) which operate continuously for 72 hours.

An FMS is composed of several machining/turning centers, work piece loading and unloading equipment (industrial robots), or material-handling equipment (conveyors/automated guided vehicles (AGV)), and possibly automated warehouses [8]. This enables unmanned production for multiple parts.

In the last few years, most large manufacturing firms have been very conscious of computer-Integrated manufacturing (CIM), which enables and requires multiple-product, small-batch production with short lead-times from the order receipt to shipment. CIM is a system integration of three different activities by computers: computer- aided manufacturing (CAM), computer-aided design (CAD), and computer-aided management [9].

- **Computer-Aided Manufacturing (CAM)** is concerned with the automated flow of materials for procurement and conversion of raw materials into products, distribution, inventory, and sales.
- **Computer-Aided Design (CAD)** involves the automated flow of technical information for R&D, product design, process design, and layout design.
- **Computer-Aided Management** deals with the automated flow of managerial information for production planning, sales planning, production scheduling, production control, quality control and cost control.

A unified combination of CAM, CAD, and computer-aided management by computers through a common database system is recognizable as a CIM.

Today textile industries in Japan are declining because of the strong competition from NIES with much lower wage rates. These same industries are trying to produce a variety of textile products of the highest quality by installing automatic machines to overcome this challenging situation. It should be noted that value-aided productivity increases with the increase in capital investment per capita of direct labor, hence factory automation raises labor productivity.

Japanese manufacturing industries now pay much attention to technological innovation for both products and processes. This continual improvement of their technologies is an important part of what might be called the 'management of technologies'.

4.2. Market-Oriented Manufacturing

Early this century the US introduced a function called 'marketing' [1]. This activity became widely known in Japan in the 1950s as the Japanese economy grew. This market-oriented manufacturing phenomenon has enabled firms to increase their market share relative to their competitors [3].

For example, one of Japan's beer breweries, Asahi, introduced a new type of 'Dry beer' in 1986. Its market share increased from 12.9% in 1987 to 20.6% in 1988 and further to 24.9% in 1989. Automobile manufacturers in Japan produced 85 models in 1969 compared to 420 models in 1986. The numbers produced per model dropped from 32 000 to 18 000. Toyota's bestselling car, the Corolla', now has 48 920 varieties in different specifications, colors, inside equipment, etc. Since this type of car is manufactured at the rate of 1500 a day, no one kind is produced even during one month [2].

Increasing the number of car models in Japan has caused the life cycle to remain at less than two years in the past ten years. This marketing concept introduced two main concepts: multiple-product, small-batch production, and diversification [6].

4.2.1 Multiple-product, small-batch production

Since consumers want products that are of high quality, specialized, fashionable, and different from others, product life cycles are rapidly decreasing. For example, Ford's T-model cars were produced for 20 years, during which time (1908-27) almost 15 million cars were made.

In contrast, a recent hit product in Japan - Minolta's automatic focus camera (a-7000) - lasted only three years, supplying three million. The average life cycle of an air conditioner is one year, that of Japanese word-processors only three months.

Hence manufacturing industries aggressively try to manufacture and supply a huge variety of products. Therefore, 'multiple-product, small-batch production' is now common. Almost 85% of Japanese products and 75% of US manufactured products fit this category.

In the US and the EC, the number of car models increased slightly to/ or around 40, and the life cycle increased from 2.5 to 5years. Hence the development period must be short in Japan, 46 months as compared to 60 months in the US and 57 months in the EC [1].

To decrease the time from design to market, 'concurrent engineering' is now employed to process various phases such as product specification, product design, process design, manufacturing equipment installation and set-up, and others in parallel. This activity is especially strong in Japan due to cooperative and consensus decision-making which is considered a special feature in Japanese management.

Multiple-product, small-batch production supplies a variety of products which consumers want, but consumers tend to dispose of still usable products. Manufacturers and consumers should be concerned with the morality of such a 'throw-away society'. Product designers, for example, are now beginning to become responsible for designing each product for its entire life cycle - integrating design, manufacture, sales, consumption, and disposal. Design cost is only 5% in the total cost; however, its influence on the future amounts to 75% (Ford Motor Company) [14].

4.2.2 Diversification

This is also known as 'economy of scope'. It is a philosophy that seeks to multiply the number of new and different kinds of products and/or services to be produced by a firm. It is now considered to be a top-level corporate strategy in Japan, with more than half of companies indicating an intent to diversify their product line. By producing and selling a number of products rather than a single product, the total production cost becomes less [7].

A recent example of diversification is found in the Canon Corporation. This company originally manufactured cameras, but of total sales in 1988 (about US$15 billion), camera sales were only 18%; other products were duplicators at a 31% share, and electronic office equipment at a 45% share. It is said that a firm's average life is 30 years. To survive this situation, firms try to restructure their organizations [11].

4.3 Global Manufacturing

This is also known as 'international production'. It was performed as early as 1865, when a German chemical maker, Bayer, merged with an American aniline company. In 1867 the US sewing machine company Singer constructed its factory in the UK. In 1909, a Japanese cotton maker constructed its factory in Shanghai. China [13].

4.3.1 Multinational Enterprises

This is the tendency of giant corporations to established their production/ manufacturing sites in foreign countries, mainly seeking low labor costs. As the trade imbalance between Japan and other countries increased, Japanese manufacturing firms invested much in East Asia, Europe, and the US.

An example shows that in 1988 a certain automobile manufacturing company made engines in Japan, chassis in Thailand, doors in Malaysia, transmissions in the Philippines, wheels in Australia. car radios in Singapore, assembled the cars in Thailand, and then exported them to Canada.

Japan's direct investment in foreign countries has rapidly increased in these years; i.e. in 1987 manufacturers invested $8 billion, almost double the previous year's investment. In 1989 Japan's foreign investment was $44.1 billion; its total investment amounted to $154.4 billion, exceeding Germany. This amount is two fifths of the US foreign direct investment, and four fifths of that of the UK Japan is third in the world. Furthermore, the percentage of total Japanese production in foreign countries is merely 4.8%, while that of the US is 16.2%.

International production or global manufacturing occasionally tends to 'deindustrialization', which means there is a decline of the domestic manufacturing sector. Japan is not in a 'deindustrialization' phase as much as the US.

Japan still keeps almost 35% of its labor force in secondary industries, including manufacturing firms, compared to 27% in the US.

5.0 MANUFACTURING CHALLENGE

Manufacturing has made the changes to come out as a major success story of the economy. That is often not recognized – too many still think of manufacturing wrongly as fixed in an epoch of heavy engineering and in decline [15, 16]. As fallow some struggles which manufacturers faced have discussed:

5.1 Intangible investment

Manufacturers in the developed economies are increasingly recognizing the importance of investing in highly productive/automated, highly flexible, low variable cost and significant information investment which calls intangibles or knowledge assets to utilize existing areas of comparative advantage in other sectors. These include design and other aspects of product development; software; brand-building; training; using highly flexible automated systems, and improvements to business processes. Such asset boosts firms' competitiveness and enables products to meet customer changing needs. The Government's role is to ensure companies have the right incentives and information to invest in intangibles, including an effective international system of official document protection.

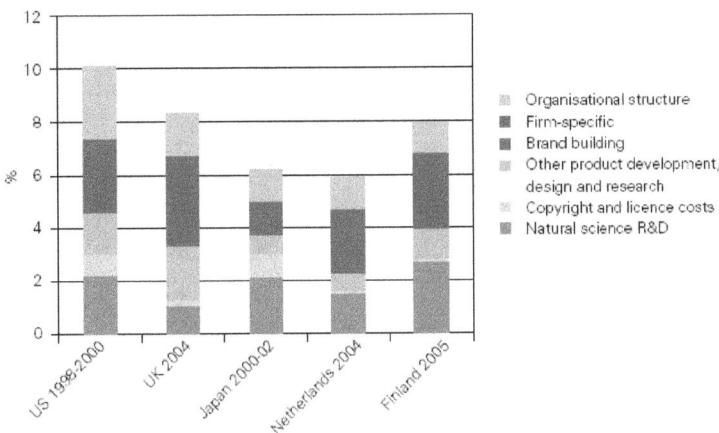

Figure 3: Investment in five OECD Countries, Percentage of GDP [17]

5.2 People and Skills

The increasing prominence of low wage economies in the global labor market and developments in technology has increased the importance of developed economies raising skill levels in their manufacturing workforces. Companies require employees with specialist high-level science, technology, engineering and mathematics skills, and a common set of soft skills enabling people to work across disciplines. The government is increasing a higher level skills policy that recognizes the significance of high level skills for modernization. Strong management and leadership is also vital for the operation of global value chains and making the most effective use of the skills of the labor force to deliver high value added products and services.

5.3 Reformation

In order to come to terms fully with investment reduction, faster product life cycles, global recession, "lifetime employment guarantees" and productivity overhang reformation is needed. So that restructured company will theoretically be leaner, more efficient, better organized, and better focused on its core business with a revised strategic and financial plan.

5.4 Poor manufacturing performance

Poor manufacturing performance has seen a power shift to retailers. As a result, industry is also looking for ways to access a greater fraction of the value-added chain (direct distribution, retailing, direct mail, etc.).

5.5 A global market environment

Unavoidably faster rates of change and more fragmentation which caused by global market environment driving companies to cope with the increasingly demanding requirement of make to order but to a short cycle.

As with all customer/supplier relationships, global relationships bring fresh demands and require innovative solutions. If goods have to travel long distances to customers, manufacturers may have to meet sudden changes in demand from a customer without holding additional local stock. Companies also need to develop new skills to manage relationships across differing languages and cultures, as well as across different time zones. In a competitive trading environment international knowledge is vital [18]

As more firms globalize, government may need to help businesses participate in global markets. SMEs in particular face barriers in accessing global value chains in high growth new emerging markets, for example in developing the skills and the collaborations needed to win access to global supply networks [18]. This is an important policy issue because successful specialization in global markets will improve national productivity and lead to the development of new comparative advantages.

To take a significant role in new international structures of value creation - where traditional definitions of manufacturing emphasize production of complete products in single locations, no longer apply- international fragment production has enabled for manufacturers.

6.0 STRATEGIES

The following strategies support manufacturers to meet these challenges and seize the new opportunities they are creating. These new strategies set out a framework that will inform a dynamic process of developing and implementing current and future policies and programmes for manufacturing

6.1 Macroeconomic Stability

Allowing businesses to plan for the long-term Investment – supporting investment in capital equipment and processes, leading edge technology, skills development, and Research and Development. It is for government to set the macroeconomic framework and create structures that will ensure continuing stability and credibility. Government needs to communicate well with other stakeholders to maintain awareness of the impact of its decisions on the real economy. It is for industry to take advantage of the greater certainty which the government's framework offers to plan and invest for the future.

6.2 Investment

Across manufacturing, all sectors have an opportunity to slim the productivity gap with their competitors by increasing investment in new technology, new products and advanced processes. Investment in capital equipment, leading edge technology, and the development of skills, creates a virtuous circle, which drives up performance. In the private sector, it's preferred for business, and financial investors, to make a decision when and where to invest. In some cases there is a need for a greater understanding on the part of companies of the benefits of investing in new products and processes in order to ensure a firm's continued competitiveness in the future. However, when, for whatever reason, the market does not function properly in providing the necessary capital for investment, there is a role for government to work with the grain of the market to help it function more effectively. Manufacturing industry is becoming increasingly internationally mobile, reinforcing the government's important role to provide an overall business climate that is helpful to firms deciding to invest.

6.3 Science and Innovation

Innovation is a key catalyst for growth, but levels of innovation have historically been low by international standards in most of developed and undeveloped countries. The aim is to raise manufacturing innovation performance, by making the best use of the excellent science base, by utilizing technology from a range of sources, and by demonstrating the profits which follow to innovative companies. Both technological and non-technological aspects of innovation are keys to success. In particular, investment, skills and best practice, and close attention to customer needs, are essential if companies are to innovate successfully. With this preamble, government has a key role in ensuring that business is able to exploit results of our science base effectively. Knowledge transfer activities can overcome information failures between business and the science base. The government also has an important role in fostering innovation for example in encouraging R&D association and knowledge sharing thus enabling individual companies to capture knowledge spin-off from each other's research, and collectively to enjoy the benefits of economies of scale and scope in innovation.

6.4 Best Practice

Adoption of best practice implies a culture of continuous improvement. Taken as a whole, manufacturers can increase their competitiveness considerably by adoption of world-class practices. It is clearly for business to recognize and adopt best practice. But the Government can play an important role in coordinating the flow of information on best practice, working closely with business networks and industry forums. Evidence shows that relatively small investment on these activities yields high returns to individual companies and to the economy as a whole. Trade Unions and employers can work in partnership to ensure best practice in management and workplace practices.

6.5 Skills and Education

Improved skills across the workforce and the creation of a system that reflects the needs of individuals and employers are essential for the completion of the government's productivity and social.

Business has a major responsibility for developing its own workforce to meet both immediate needs and the demands of industrial change. So manufacturers must work closely to maximize the contribution of both higher and further education to the productivity agenda.

6.6 Modern Infrastructure

A company needs a modern, efficient public infrastructure to enable business to reduce costs, increase efficiency and improve its competitiveness. This is a major challenge given decades of under-investment. The transport system is of central importance, together with a thriving broadband market due to the growing importance of e-business in all sectors.

6.7 Right Market Framework

Right market framework provides the helpful business environment that manufacturing needs to compete globally. This requires competitive and dynamic markets and motivated, well informed and confident participants – business, consumers, employees and investors.

7.0   CONCLUSION

In this article different types of manufacturing systems has been mentioned and recent trends in manufacturing has been discussed. Nevertheless manufacturing was and continues to be a fundamental and useful activity in human history.

A survey of recent articles in this subject and content analysis has shown that manufacturers should apply and come up with new trend such as technological innovation, mass customization and globalization in order to survive among their rivals in our highly civilized society. With applying innovation both in design of product and manufacturing processes, manufacturers have scored point against their competitors by meeting their customers' demands and gaining dominance share of the world market.

By considering different cases of market oriented and global manufacturing, it shows that there is significant increase in production rate with aim of seeking low labor cost in different firms. However, applying these new methods leads manufacturers to face with different challenges that require long-term planning, skill development, right market framework and good knowledge of new methods.

REFERENCES

[1] BERR Economics Paper No. 2: 'Five Dynamics of Change in Global Manufacturing' – *Underpinning Economic Analysis* www.berr.gov.uk/fi les/fi le47663.pdf

[2] Daniel J. Ryan, Trends in market power and productivity growth rates in US and Japanese manufacturing, Economics Department, Temple University, Philadelphia, PA 19122, USA

[3] Historical Statistics, 1960-1987. OECD. 1989, p. 59.

[4] H.-P. Wiendahl; H.A. ElMaraghy Changeable Manufacturing - Classification, Design and Operation *CIRP Annals - Manufacturing Technology, Volume 56, Issue 2, 2007, Pages 783-809*

[5] Intellectual Assets and Value Creation, Synthesis report' OCED Science Technology and Industry Directorate, 2008. p 13

[6] J.G. Abegglen, The Japanese Factory. Free Press, 1958

[7] K Hitomi, Manufacturing Systems Engineering. Taylor & Francis, London, 1979

[8] K Hitomi, Manufacturing systems engineering: the concept, its context and the state of the art. Inter- national Journal of Computer Integrated Manufacturing, 3(5) (1990) 275-288.

[9] K Hitomi. Non-mass, multi-product, small-sized production: the state of the art. Technovation, 9(4) (1989) 357-369

[10] K. Hitomi. Steps toward CIM: system integration of computer-aided design, computer-aided manufacturing and computer-aided management. Japanese. Journal of Advanced Automation Technology, 2(4) (1991) 7-11.

[11] K Hitomi, Strategic integrated manufacturing systems: the concept and structure. International Journal of Production Economics, 25(1-3) (1991) 5-12.

[12] K Hitomi, The Japanese way of manufacturing and production management. Technovation, 3(1) (1985) 49-55.

[13] K Ishikawa, What is Total Quality Control? The Japanese Way. Prentice - Hall, 1985.

[14] K. Sakai, The feudal world of Japanese manufactuting. Harvard Business Review, 68(6) (1990) 38-47.

[15] L.P. Alford, Laws of Management Applied to Manufacturing. Ronald Press, 1928, p. 135

[16] Overby, Design for the entire life cycle: a new paradigm? AASEE Annual Conference Proceedings, ASEE, 1990, pp. 552-562.

[17] Simeon Nichter, Lara Goldmark Small Firm Growth in Developing Countries, World Development, In Press, Corrected Proof, Available online 26 June 2009.

[18] Y. Monden, Toyota Production System. Industrial Engineering and Management Press, 1983

*****

Indicators and Models of Sustainable Development

Amin Teyfouri [1a], *Parisa Keyhani Boroojeni*[2b]
[1] *Engineering Faculty, UPM, Malaysia*
[2] *Engineering Faculty, UPM, Malaysia*

Academic Editor: *Shahryar Sorooshian*

ABSTRACT: This paper takes account of the development of three indicators of sustainable manufacturing including environment stewardship, economic, and social well-being. We focused on plant waste depletion and associated sustainability in regard to environmental matters in Industrial Ecology (IE), Life Cycle Assessment (LCA) and Manufacturing System (MS) modeling. The study highlights the importance of bilateral relations between techno sphere and ecosphere. We conclude the paper by investigating characters of sustainable manufacturing for improving waste, use of material, and consumption of energy during production via some models.

Keywords: Sustainable development indicators; Industrial Ecology (IE); Life Cycle Assessment (LCA); Manufacturing System (MS)

1.0 INTRODUCTION

It may indeed be true to say that sustainable manufacturing plays significant role in modern industrialization. As a matter of fact sustainability is meeting the demands of human being while do not disturb the demands of next generation [1]. As a general rule of thumb, manufacturing sectors are the principal client of the energy and materials [2]. Thus meeting the growing demand for resources of energy and material is an undeniable problem. According to the environmental activists' opinion, those who perceive green rules would merit more than the ones who do not obey environmental issues. The ultimate target of industrial systems is employing technology and procedure to revolutionize inputs for achieving a profitable output. Promotion manufacturing products and manufacturing processes that can cut environmental impacts and enhance social and economic benefits is an aim for every manufacturer. Sustainable manufacturing has noteworthy impact on company's new products and competitiveness [3] in today's world since the customer's trend towards sustainable strategies has been urged. Section 2 reviews a collection of structure types of manufacturing. Section 3 provides the indicators of sustainability in manufacturing. Section 4 provides an illustration of model. Section 5 depicts the meaning and a perspective of SM. Section 6 Suggests the LCA model and its feedbacks then Section 7 presents summarizes this work.

2.0 PROGRESSION OF SUSTAINABLE MANUFACTURING

It's clear that these days the structure of manufacturing is being improved dramatically. During three decades recent research provides an overview of four movements of manufacturing. It involves Traditional Manufacturing, Lean Manufacturing, Green Manufacturing, and finally Sustainable Manufacturing. Innovation elements of 6R plays significant role in these cycles; in fact, 6R involves Remanufacture, Redesign, Recover, Recycle, Reuse, and Reduce [4].

Remanufacture: It involves re-processing of the used-products for a new ones so that it doesn't lose its performance.

Redesign: Create product more sustainable.

Recover: It involves gathering the products (at the last step), stripping, organizing, and cleansing in order to using later.

Recycle: The process of changing waste material into new material for using productive cycle

Reuse: It has been mentioned to reuse goods or its mechanisms, before the second life-cycle that are resulted in reduce of usage new raw material for process of producing.

Reduce: It embraces correct use of energy and materials in manufacturing and declining waste during each stage of production.

3.0 SUSTAINABLE DEVELOPMENT INDICATORS

The world has been making an effort to acquiring sustainable development via new approaches. The indicators of sustainable development cover society, environment, and economic in terms of quality, cost, and a process during manufacturing. Directly, many approaches have emphasized on more environment in comparison with another factors in sustainable development due to its importance that is more than society and economic.

3.1. Indicators of environmental stewardship

In fact for environment stewardship can classify to effect of emissions, amount used of resource, pollution, and the natural habitat conservation

3.1.1 Emissions

Emissions are classified to effluent, solid waste emission, air emission, and waste energy emission [4]. Indicators in the emission subgroup are indicated what society or cycle discharges during the production

3.1.2 Amount used of resource

This topic is classified to consumption of water, consumption of material, consumption of energy, and plot use for cycle or process. The use of material subgroup is included indicators of total material use, virgin material use, recycled material use, reused materials, remanufactured materials, and other material uses [5]. The energy use includes overall energy use and keeping energy

3.1.3. Pollution

Pollution is included to dangerous substances, Green House Gases (GHG), ozone-depleting gases, and the like that are harmful for nature [5].

3.1.4. Natural habitat conservation:

This is classified to: biodiversity, habitat management, and conservation. This subcategory is required to reflect impacts on wildlife species with the habitat in the place of living.

Today's many companies have industrialized assessment equipment for concentrating the environmental aspect of sustainability such as the Eco-Indicator and Life-Cycle Assessment [4]. The future plan for the environment stewardship dimension ought to be on the additional assessment of the current indicators, which have several worries and uncertainties in their files collection approaches.

3.2. Economic development indicators

Economic growth indicators contain amount profits, costs, and investment for a business. Actually, economies in modern world concentrate on materials and energy. And the large number of these materials become waste relatively rapidly (only 6 % of this active material is alive in tough properties; the other 94% is converted into waste inside a few months of existence removed [1].

- Proceeds indicators are used for determining the profit acquired by the organization

- Cost indicators are used for evaluating costs from producing, including material achievement, production, productive delivery to client, and end-of-service-life product management. These factors are recognized from many manufacturers during business accounting and life-cycle estimating.
- Investment indicators. This factor is used to calculate the results of the effects from usual and eco-friendly outlay, which reaches the economic health in a company. It's noted that usual investment can help to business and public growth of the organization that are included investment clearly that inspire environmentally friendly investments. Many of the indicators inside the economic growth aspect are piece of long starting cost accounting, pros of investigation, life-cycle costing, and risk management approaches and have been broadly used and known by several of organizations. The advance decay of actions and processes inside the organization can provide more indicators indoor the economic development measurement.

3.3. Social well-being indicators

Social well-being indicators measure social aspect of manufacturing processes and industrial products through usual health and safety performs, growth management, and human rights by an organization. The social well-being aspect erection indicates three basic dimensions. Employees, customers, and the close community are rightly or wrongly affected by the movements of an organization, and taking into account the effects of communally sustainable operation and general organization sustainability.

3.3.1. Employee

It includes general health and safety of employees, their specialized development, and happiness in an organization. In fact these indicators in the employee are essential for sustainable manufacturing due to human rights problems, but also strong connection between employee and product quality.

3.3.2 Customer

Customer indicators include the health and safety effects from manufacturing and product, client satisfaction from processes and yields. The customer subgroup covers indicators that indicate the skills of the organization to fulfill demands of customers. The customer satisfaction indicators are necessary to amount customer satisfaction and well-being as those are can be important influences for creating an organization or business.

3.3.3 Community indicators

It is associated with an organization's movements. Subgroups includes responsibility for products, justice (Fairness, equity, human, rights), and community development programs. Community measures are healthy connection for organization's community.

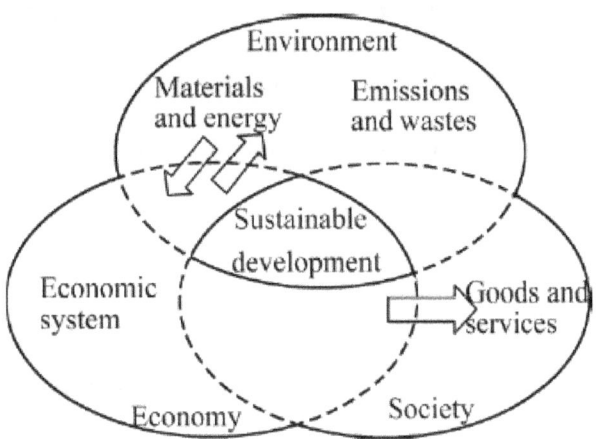

Figure1. A model of sustainable development [6]

Actually, there is not still total explanation for sustainable development, many reach agreement that sustainable development is about satisfying social, environmental and economic goals. Therefore, there is no doubt that the statistics of three dimensions of sustainability require to collect and organizing by industrial methods. It is logical that the factors of goods and services are one prominent technique to reaching sustainable development; in other words, the correct method of using resources and energy in manufacturing is the best possible way for creating the fewest amounts of waste and emissions. Another key element is use of renewable resources because the decrease of cost in each stage. While sustainable economic development will necessitate to changing the type and amount of resources in manufacturing.

4.0 INDUSTRIAL ECOLOGY

Industrial Ecology (IE) is a well-known mode with a large number of cases at inter-enterprise yield. It is related to the study of connections and contacts between industrial systems and natural systems. In a factory, IE integrates three major parts of a manufacturing system which are material, energy and waste .These three parameters are known by MEW [7] .They contain manufacturing operations, plant and facilities.IE is categorized in three types for flows of resource:

"Type I": In linear type, the system highly depends on outside resources as it accepts that the environment has infinite amount of source for providing its inputs besides absorbing waste of production.

"Type II": In quasi-cyclic type, the system declines its demand of sources for input and waste of production as there is a specific degree of recycling of material and energy which are our resources.

"Type III": The system has the lowest dependency. Cyclic resource flow illustrates utmost level of cycling which is done in closed-loop and it merely needs energy (solar) as input and means self-containment. [8]

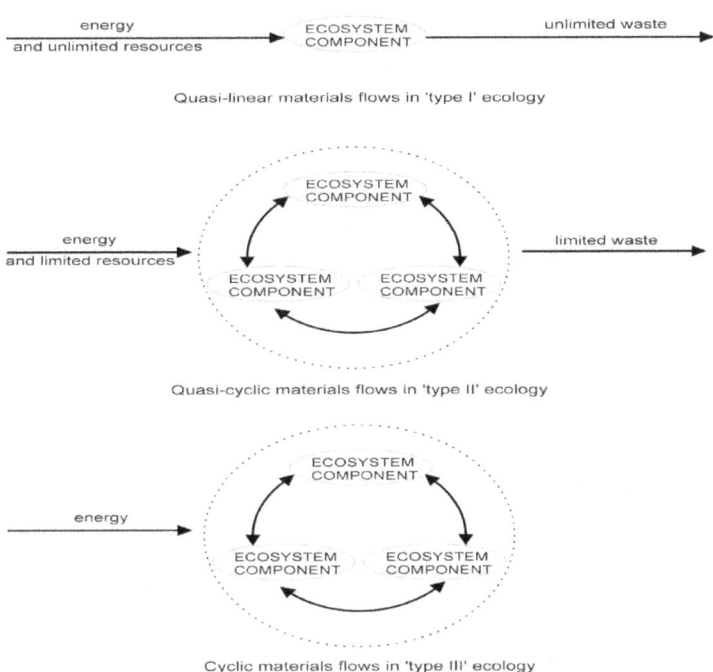

Figure2. [9]

Industrial ecology is followed by three levels: the micro level that is the firms, macro level and eco-industrial parks which contain more chances [10] . In an ideal IE the industrial systems is established on variation interdependency, assistance and recycling that the industrial systems is found as a local collaboration system which admires natural resources and goes after substitution local resources and reproduction. [11] On the whole, IE tries to transform waste of outputs into reusable inputs which means transposition of technosphere to ecosphere .nutrients. Moreover, IE suggests the use of local sources to cut the expenses that are incurred thorough imports and exports and finally decreases the impact of them on the environment.

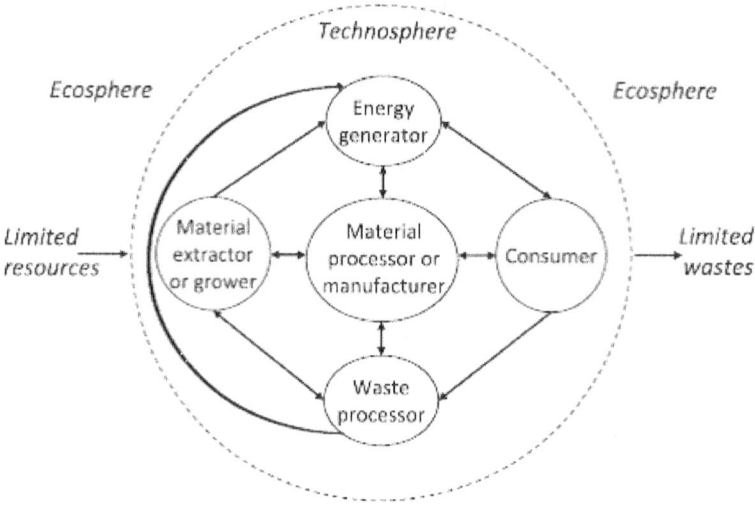

Figure3. [7]

5.0 SUSTAINABLE MANUFACTURING (SM)

The environment is a major issue in the contemporary world. Owing to raising environmental awareness, industrial sectors implement new policies to use materials which are biodegradable or eco-friendly, renewable or clean energy and sophisticated technology to devise and settle new machines which all can lead to reduction in economy (higher profit) too. Broadly speaking, there are three stages in every manufacturing cycle that we are facing the problems which can harm the environment and should be handled. It starts by choosing the proper and adequate recourses then follows by electing the appropriate technics and processes to gain the thorough products and ends by closed-loop (roundput) which consists of reprocessing, reuse and recycling [7] . In the past SM was applied in the name of IE, but nowadays it appears itself as a separate issue. Recently SM concentrate on remanufacturing and reuse subjects for maintaining the output at the end of process and reuse them again as input. Environment and technology are two major issues in SM and they have some contrasts since more sophisticated and easer lives come from technology and this technology may in danger the environment and ecosystem which is a growing alarm for future generations. Thus

SM make an attempt to pursue several rules such as utilize less: work towards the more efficient use of all natural resources, reuse waste outputs as new input (remanufacturing), replace some biodegradable inputs by non-biodegradable ones. The SM ideas are being employed to product design rather than design of manufacturing systems [12]. There are just a few researches that concentrate on enhancing the particular manufacturing systems especially in the assembly phase and the part production [13].

6.0 LIFE CYCLE ASSESSMENT

Life Cycle Assessment (LCA) is generally known as a basic tool to find how each product can influence an ecosystem during its life cycle. These days it is being the center of attention among society, industries and also governors Analysis of lifecycle costing is a highlight tool in LCA that contains the biosphere and green policies. According to the examination the decision that comes from teams is more reliable than decision which is made by individuals especially by virtual ones [14].Life Cycle Assessment tries to figure out the whole effects of economic (profit) and environment (plant) in relation to carbon foot print, energy, material and etc. [4]

Being time consuming in quantity and requiring certain data that is usually unavailable are the main problems of LCA [15]. Sustainable product development and manufacturing by considering environmental requirements LCA are mostly applied on product and supply chain levels than plant since the environmental problems are generally seen in outputs [16] .In LCA we should consider the variation

n of life cycle in products. It would be shown the impact of products in third world while LCA is applied to illustrate the raw material [17]resources and the final destination of the wastes of product [11].

7.0 CONCLUSION

It recommends take an action to minimize the harmful effects of manufacturing on our habitat by suggesting some practical models.

This study concentrates on resource flows to explore the place that the waste outputs can be used as the inputs of a new process to preserve the resources as much as possible. It has been urged to implement the closed-loop in each step of a product life instead of the end-of-life. Beyond a shadow of doubt, protecting and conserving the environment will be a critical subject in the future owing to the lack of fossil fuels as a major source of energy and raising demand for material hence we require the new way of thinking to develop sustainable manufacturing.

REFERENCE

[1] S. L. R. D. D. approachMargot J. Hutchins∗, "Journal of Manufacturing Systems," *Understanding life cycle social impacts in manufacturing:A processed-based approach,* 2013.

[2] F. M. A. P. K. C. M. N. Amir Rashid*, "Journal of Cleaner Production," *Resource Conservative Manufacturing: an essential change in business,* vol. 57, p. Journal of Cleaner Production, 2013.

[3] D. C. E.W.T Ngai, "Energy andutilitymanagementmaturitymodelforsustainable," 2012.

[4] F. B. O. D. J. I. J. A.D. Jayal, "CIRP Journal of Manufacturing Science and Technology," *Sustainable manufacturing: Modeling and optimization challenges at the product,,* vol. 2, p. 144–152, 2010.

[5] J. C. ,. P. S. S. C. F. Che B. Jounga, "Categorization of indicators for sustainable manufacturing," *Ecological Indicators,* vol. 24, p. 148–157, 2012.

[6] Csiro, "Material and sustainable development," *progress in natural science:material industrial,* vol. 20, pp. 16-29, 2010.

[7] P. B. S. E. A. L. M. Despeisse*, "Industrial ecology at factory level e a conceptual model," *Journal of Cleaner Production,* vol. 31, pp. 30-39, 2012.

[8] J. Korhonen, "Journal of Cleaner Production," *Four ecosystem principles for an industrial ecosystem,* vol. 9, p. 253–259, 2001.

[9] M. B. ,. A. M. ,. A. D. Ewa Liwarska-Bizukojc, "Journal of Cleaner Production," *The conceptual model of an eco-industrial park based upon,* vol. 17, p. 732–741, 2009.

[10] B. H. Roberts, "Journal of Cleaner Production," *The application of industrial ecology principles and planning,* vol. 12, p. 997–1010, 2004.

[11] J. Korhonen, "Journal of Cleaner Production," *Industrial ecology in the strategic sustainable development model:,* vol. 12, p. 809–823, 2004.

[12] L. W. Z.M. Bia, "Journal of Manufacturing Systems," *Optimization of machining processes from the perspective of energy,* vol. 31, p. 420– 428, 2012.

[13] P. Leigh Smith, "Int. J.ProductionEconomics," *Steps towardssustainablemanufacturingthroughmodellingmaterial,,* vol. 140, p. 227–238, 2012.

[14] G. S. c. S. c. M. c. B. M. Gaussin a, "Int. J.ProductionEconomics," *Assessingtheenvironmentalfootprintofmanufacturedproducts:Asurveyof.*

[15] S. K. M. S. H. Kaebernick, "Robotics and Computer Integrated Manufacturing," *Sustainable product development and manufacturing by,* vol. 19, p. 461–468, 2003.

[16] H. Winkler, "CIRP Journal of Manufacturing Science and Technology," *Closed-loop production systems—A sustainable supply chain approach,* vol. 4, p. 243–246, 2011.

[17] Q. Z. ,. D. D. Chris Yuan, "CIRP Annals - Manufacturing Technology," *A three dimensional system approach for environmentally sustainable,* vol. 61, p. 39–42, 2012.

[18] J. C.-A. K. R. H. Hao Zhang, "Journal of Manufacturing Systems," *A conceptual model for assisting sustainable manufacturing through systemdynamics,* 2013.

*****

Review on Supply Chain Management: 4th Party Logistic

Milad Ahmadi

UPM University, Serdang, Malaysia

Academic Editor: *Amin Teyfouri*

ABSTRACT: In this review paper, a brief of current problem of the internal logistic department is address. Among the problems are logistic department would have to collaborate with many logistic providers depending on availability of service area, issues in returning goods and recyclable packaging and the complicated arrangement with the customer on the availability of returnable goods. The relationship of logistic in supply chain and its complexity is explained in the introduction. A general development of logistic provider is also defined in this section from 1PL to 4PL provider. Then this paper continues to explain in details of 4PL services such as the definition, roles, categories of 4PL, the current trend of 4PL application, what information is needed and how the system flows the information in 4PL and the benefits of using a 4PL provider. This paper will also brief on how to select a 4PL providers, the importance of performance measurement and the risk associated with 4PL. In conclusion, the writer will make a connection of the 4PL and how it can improve the current condition of problems in the internal logistic department.

Keywords: supply chain management; 4th party logistic

PROBLEM STATEMENT

Logistic plays an important role in supply chain management. Proper logistic management ensures that customer receive the goods and services at the specified location, time and quantity. Failure to meet the customer specification and requirement will create dissatisfaction and possibility of losing future projects. A marketing department can easily influence and give prefect image and condition of an organization to a customer before securing a project, however, in sustaining and impressing long lasting relationship, among the first factor that will be measured by a customer is the delivery of the goods and services other than the quality of it. Logistic will give an overall picture of the organization actual operation of planning, production and quality.

In today's supply chain environment, many organizations are relying on their own logistic department to plan and coordinate logistic of suppliers and customers. To deliver products, organization will either directly deliver the product with their own transportation; subcontract the logistic to a third party logistic or if delivery to certain customer is seldom, a one off services from a third party logistic (3PL) provider is made. One organization may have much type of logistic services as not all logistic providers are able to cater requests from the organization. For example, for customers who is located in the north, logistic provider A will be used and for customer at the south, logistic provider B will be used, for deliveries within 100km radius, organization would use their own transportation and for customer located at overseas, logistic provider C and D will be used depending on who provides the services at which area. Other than that, logistic personnel would have to make different arrangement for goods and recyclable packaging that is to be returned from customer. Many logistic providers are unwilling to support this activity as the return of goods and recyclable packaging is in a very small quantity and there are not many organization requests for arrangement of returnable parts. Not only that even the organizations have issues especially related to paying the logistics provider with a full truck cost for a small quantity of non-value added services and logistic planners would have to collaborate with customer of the readiness of the parts. So with this many type of logistic provider services, and the upstream and the downstream flow of the goods, internal logistic department would have to handle a truly complex supply chain situation. Most small and medium organization would only have one or two personnel in charge if the logistic department and would be a handful to them.

INTRODUCTION

LOGISTIC AND SUPPLY CHAIN MANAGEMENT (SCM)

SCM as the process of planning, implementing and controlling the operations of the supply chain with the purpose to satisfy customer requirements as efficiently as possible [14]. SCM is a systemic, strategic coordination of traditional business functions and the tactics across these business functions within a particular company and across businesses within the supply chain for the purpose of improving the long term performance of the individual companies and the supply chain as a whole. In a supply chain, there is a flow of product or services, finance, and information from top to down and reverse.

There are three degrees of supply chain complexity which is the direct supply chain: consists of supplier, manufacturer and customer, extended supply chain: consist of supplier's supplier such as second tier supplier, first tier supplier, manufacturer and customer, the customer' customer such as the end user and lastly the ultimate supply chain: consists of all the organization involve [1].

Direct Supply Chain

Supplier ⟷ Manufacturer ⟶ Customer

Extended Supply Chain

Supplier Tier 2 ⟷ Supplier Tier 1 ⟷ Manufacturer ⟷ Customer
End User

Ultimate Supply Chain [1]

Logistic Provider   Logistic Provider

Supplier's Supplier ⟷ Supplier Tier 1 ⟶ Manufacturer ⟷ Customer ⟷ Customer's Customer

Financial Provider   Financial Provider

The systemic coordination and organization unifies the internal and external operation as a whole to mainly create customer focus which in the end targets the customer satisfaction. This is achieved through integrating the behavior of each entity by sharing information, process, risk, reward and maintaining long term relationship.

As described in the ultimate supply chain, logistic provider plays a great role in delivering products from supplier to manufacturer and from manufacturer to customer. In this case, manufacturer and supplier choose to outsource the logistic service to the availability of logistic provider. It is different in the case of direct supply chain and extended supply chain, manufacturer and supplier manage the logistic on their own. Define by [12] logistic is part of supply chain process that plans, implement, and control the efficient flow of storage, goods, and related information from the point of origin to the point of consumption order to meet customer requirement.

Other than the downstream flow of goods and services, there is also the existence of reverse logistic. Reverse logistic is the process of planning, implementing and controlling the efficient, effective inbound flow and storage of secondary goods and related information opposite to the traditional supply chain direction for the purpose of recovering value or proper disposal [10] such as returning damaged goods [5], returnable packaging or excess goods which is reach the end of life status or which was wrongly delivered by supplier or manufacturer to their respective customers. Reverse logistic is part of the after sales service and has become important as the increase of as the increase of online purchases, stricter environmental regulations, higher quality standards, more lenient return policy and more demanding customer [5] creates a more competitive edge in the industry. The question is now how SCM players can effectively and efficiently manage the transportation of goods back to manufacturer and supplier?

LAYERS OF LOGISTIC

There are several layers of logistic which consist of first party logistic (1PL), second party logistic (2PL), third party logistic (3PL) and fourth party logistic (4PL).

According to [15], 1PL are direct logistic services from manufacturer to customer or supplier to manufacturer, 2PL are the carriers which provides transportation services such as maritime shipping company, rail operator or trucking company, 3PL are services of receiving, holds and transport goods with using their own assets such as transport and warehouses while 4PL are the single interface for an organization to integrate its logistic functions without having any assets. This would mean, a 4PL could be using multiple 3PL providers in their services.

The relationship of 3PL and 4PL are very close, however in below section, the difference of 3PL and 4PL are further explained.

Difference of 3PL and 4PL

Table 1: Difference between 3PL and 4PL sourced from [3][5]

No	Third Party Logistic (3PL)	Fourth Party Logistic (4PL)
1	Asset based	Non asset based
2	Partly accountable with internal resources or other 3PL	Full accountability on all users outsourced operation.
3	Focuses on logistic function such as warehousing, distribution, tracking deliveries, specific packaging.	Focuses on logistic, supply and demand chain integration by using design and planning capabilities with advance IT solution.
4	Performance measurement based upon cost	Performance measurement based upon value created within client's organization
5	Request based	Requires transparency and real time information
6	Organization may have more than one 3PL	Single interface between organization and multiple 3PL

4PL INTRODUCTION

Arthur Anderson now known as Accenture Consulting in 1996 proposed a definition of 4PL as an integrator that assembles its own resources, capability and technology and those of other service providers to design and manage entire sets of complex supply chain [2][6]. 4PL is an outsourcing distribution network management of logistic services that change the phase from a single function to multifunctional supply chain It has an important role to manage the supply chain by controlling the time and place. Its services include custom consultants in managing inventory, planning of physical distribution of transport and distributes flow of information [3][6]. To provide a comprehensive successful solution and ensuring consistency in their services, 4PL will have to integrate and process all relevant knowledge in form of information, skills and information technology. These is because knowledge is becoming the only resource that is capable of offering competitive advantage and continue growth and prosperity of supply chain partners [13]. Therefore in 4PL, to link and analyze all the available information is essential as it will a enlarge the dimension of information across supply chain connection [14].

By forming strategic alliances between suppliers and their customers, 4PL is able to provide competitive edge in creating customer value and allows single point of accountability across supply and demand [3][11]. Risk of quality assurance in planning and logistic to customer for the participated supply chain supplier transferred to 4PL. Therefore, in working toward success relationship, these strategic alliances, creates closer relationship among supply chain entities and promotes greater flexibility of demand and supply uncertainty. According to research conducted Frost and Sullivan in 2005, the 4PL market as a whole is expected to witness considerable revenue growth from approximately EUR 4.7 billion in 2002 to about 13 billion by 2010.

CATEGORIES IN 4PL

There are 4 categories of 4PL providers defined and explained by [6] such as 4PL in document engineering, 4PL in logistic simulation, 4PL in subsidiaries group of logistic and 4PL pure players. Below are further explanations of these categories sited from [6]

1. 4PL in Document Engineering

   It is specialized in document engineering for large project including information, data and documents. It is designed for management and technical information in the informational chain of large project. Some examples of project are:-

   a) Writing and translating documents of manual, product specification, civilian and military standard, quality documents, plan and specification.

   b) Storage and management of flows including archiving, digitization and circulation of data and documents to all partners

   c) Develop information system of documents production and consultation, management and circulation of technical libraries, knowledge capitalization software.

   d) Analysis of logistic support such as profitability analysis, assistance in selecting and supervising subcontractors, and suppliers, technical and maintenance assistance in contracting project management.

2. 4PL in Logistic Simulation

   Consists of software publishers and mainly consult and stimulate global logistic solution such as

a) Modeling of supply chain elements of supplier, factories, warehouses, distribution networks and clients and their global integration into a project. Optimizes the capacity of warehouses and operational logistic operation and modeling performance.

b) Transform computer infrastructure and trade operation, improvement of transformation flow throughout the supply chain, improving the planning and organization of logistic in general using a global integrated information system and logistic data bases.

c) Outsource through software program which was previously developed by 4PL and also through mission facilities management to add value to the client or customer information system

3. 4PL in Subsidiaries of logistic groups

Independent companies or subsidiaries of 3PL group which offers logistic consultation and organization to various clients in diverse sectors such as:-

a) Global, collaborative and flexible solution for the supply chain for example management of logistic project, operational assistance and principle contracting in logistic information system and assignment of dedicated staff.

b) A network of partners to provide client with the benefit of good coordination of outsourced transport flows

c) Visibility and quality in management of outsourced logistic administrative process, custom engineering and procedures dedicated to industrial tracking of ISO standards service provision.

d) Continuously improvise budget control for administration of sales, reporting and dashboard key performance index, KPI.

4. 4PL Pure players

There are only a few of 4PL pure players which could design, organize and coordinate the whole of logistic, document, and regulate chain. Pure player is made of large companies which satisfy six to nine criteria of :-

a) Expertise in consulting which possessed with no physical assets related to the activity. Uses intellectual and computer resources to define and optimize supply chain, manage and synchronize all the flow of upstream and downstream supply chain.

b) Integrator of activities such deployment of staff to clients for a specified period to best accompany them

c) Transport broker for flow of goods and experts in local and international regulation, insurance law, selection of subcontractor and other service providers.

d) Transfer liability which responsibility is in their own and clients name for all contracts with service provider, suppliers and partners

e) Creates and has own international network to meet clients need with several subsidiaries of local officers.

f) Custom broker which manages custom formalities of drafting custom declaration, information of nomenclatures, established rate for duties, taxes and physical control of goods.

g) Specialized in segment which can be based on international routes, or one or more modes of transport for certain sector of several clients

h) Capable in real time tracking and traceability of good based on specific information system.

i) Has a complex computer solution dedicated to logistic to conduct and integrate combination of own capacity and outside expertise.

CURRENT TREND OF 4PL

Through the four categories of 4PL, the most popular practice sees that 4PL providers integrate the competencies of several 3PL such as UPS and e Logistic Control. The practice of 4PL in document engineering is relatively very small and need further enhancement. This is because organizations are still skeptical of sharing valuable information to an integrator which could affect their position in this competitive market. Little do they know that 4PL is a strategic alliance which leverages the skills, strategy, and technology locally and globally which might take many years to duplicate [3].

WHAT INFORMATION IS NEEDED AND HOW THE SYSTEM FLOWS THE INFORMATION IN 4PL?

4PL providers have an information platform which is design to support large scale of information and operational data from multiple sources. Among the most important information exchange between supply chains entities describe by [2] is:

a) Logistic basic information

b) Third Party Logistic (3PL) basic information

c) Customer and suppliers information interpretation

d) The appraisement of 3PL suppliers

e) Logistic project optimizing and decision making

f) Choosing traffic routes and 3 PL supplier [2]

These data is then integrated in the 4PL information system and analyzed before a decision is made. Although the existence of complex information technology and system could be programmed for decision making, an ultimate decision from an expert is essential to ensure smooth flow of the process.

BENEFITS

Through the support of information platform and system, 4PL is able to solve multiple problems from different sources [2]. For example if one organization need to deliver product at location A but then does not meet the economics of scale, the organization would still require to pay the same amount to the 3PL provider. However, if the organization decides to share its information and problem with a 4PL, and there are another organization B which could fill the truck load and have the same destination, a 4PL will be able to integrate the requirement and arrange for a 3PL services to meet both request. This then would led to sharing cost of transportation between organization A and B. Organization A and B would have no knowledge of the other requirement without the assistance of 4PL. Even if the organization decides to use the same 3PL, both would have to pay the full truck load price respectively. Consequence, both organizations is able to minimize the logistic cost. 4PL would also be able to arrange reverse logistic among its participant as even when the quantity is relatively small for delivery for one company, but through the integration, 4PL are able to cumulate goods of other supplier from the customer to ensure full economic of scale are achieved.

SELECTING A 4PL PROVIDER AND ENSURING SECURE RELATIONSHIP

Several factors are important when identifying and selecting a 4PL. A 4PL provider must be able to manage activities of multiple 3PL providers, has the experience in facilitating and able to demonstrate to manage supply chain integration and fluctuating demands, understands and the industry which the service is provided, has the ability to operate at operational, strategic and tactical level, is capable of cost control [3]. In 4PL, sharing of information and knowledge is essential; therefore it is highly important for management to choose level of alliances with entities in supply chain [8]. Management must be able to differentiate the categories in 4PL before making a decision as some information which might be important in the procedures of 4PL in ensuring success of the complex supply chain. However, this information is confidential to the organization [4].

PERFORMANCE MEASUREMENT

Performance of a 4PL companies are measured through the function of value creation within the client organization and are assessed through financial and non-financial indicators such as inventory investment and stock turns, lost sales, days out of stock, service inventory level, inventory aging, customer service perception, customer complain, cost of supply chain operation, amount and cost of expediting and effectiveness of demand forecast management [3]. It is important for service provider to align their performance measurement strategy with their client to decrease potential risk of losing client while increasing the quality of service [5]. Having a key performance index (KPI) which was established between client and a 4PL provider is crucial as it ensures that the 4PL providers are able to perform as what has agreed. Through the KPI, 4PL providers are able to develop organizational, procedural, technical and strategic capabilities to respond to four emerging requirements which are customer and end customer focus, technological adoption, relationship management and style of leadership [11]

RISK

There are several risks which could be associated with 4PL service provider alliances which are operational risk, organization loss of control and different culture practices.

Operational risk from an unsuccessful implementation of 4PL service provider and technological risk from failure of complex information technology system implemented which would then resulted in 4PL service provider not being able to meet the organization targets of service level for example, logistic failure of deliveries and not being able to prove cost reduction for the organization [9].

There are potential losses of control [9] in certain organization activities if the organization is too dependent of the services provided. 4PL does not necessarily involve in the organization operation however, it would create intense situation when certain activities are not in the organization's control and are being pressured by the 4PL [7].

Cultures from each organization are different between one another, and would create a stringent environment between both entities. Therefore, there must be a medium or organizational mechanism to assist in culture integration to ensure the success in implementing 4PL.

CONCLUSION

In conclusion, 4PL characteristics and benefits are able to assist an organization in organizing its logistic system upstream and downstream. 4PL integrates all relevant information from each organization of suppliers and customers which participate in their services and provides a win-win situation. To benefit from 4PL, an organization must make proper selection of 4PL categories, provider and ensure that KPI is established. Communication and information is crucial are these are the important tools for a 4PL to conduct their services. For 4PL providers, in order to have customer believe in their future gains in logistic compared to current practice, a 4PL providers must have a more intensive use of information technology, a pooling of resources and skilled human capital with significant intellectual capacity [6] A company can outperform rivals only if it can established a difference that it can preserved and must deliver greater value to customer to create comparable value at a lower cost [3].

REFERENCE

[1] J.T. Mentzer, W. Dewitt, J.S Keebler, S. Min, N.W. Nix, Defining Supply Chain Management, Journal of Business Logistic, Vol. 22. No.2 (2001).

[2] L. Changshi, N. Kai, Fourth Party Logistic Information Platform Based on Virtual Data Warehouse XML, Economics and Trade College, China.

[3] A. Win, The Value A 4PL provider can contribute to an organization, International Journal of Physical Distribution and Logistics Management, Vol. 38, No.9, (2008)

[4] H.Z Hessami, A. Savoji, Risk Management in Supply Chain, International Journal of Economics and Management Science, Vol.1, No.3 (2011) p 60-72.

[5] P. Patersson, T. Zantvoort, Fourth Party Logistic: A case study on performance measurement, Master Thesis, Jonkoping International Business School, Sweden (2012).

[6] L. Saglietto, Towards Classification of Fourth Party Logistic (4PL), Universal Journal of Industrial and Business Management,(2013).

[7] M.S. Sohail, A.S. Sohal, R. Millen, The state of quality in logistic: evidence from an emerging Southeast Asian nation.

[8] L.M. Jensen, Humanitarian cluster leads: lesson from 4PLs, Journal of Humanitarian Logistic and Supply Chain Management, Vol. 2, No.2 (2012)

[9] M. Falea, Outsourcing Logistic Activities, Supply Chain Management Journal (2011)

[10] M. Fleishmann, Quantitative Model for Reverse Logistic, Thesis, Erasmus Univeristy, Rotterdam (2000).

[11] Langley, C.John, M.C. Holcomb, Creating Logistic Customer Value, Journal of Business Logistic (1992)

[12] Council of Logistic Management, Oak Brooke II (1998)

[13] C. Wu, Knowledge Creation in Supply Chain, an International Journal of Supply Chain Management, (2008)

14] M. Mahajan (2012), Industrial Engineering and Production Management, Dhanpat Rai & Co, India.

[15] Information on http://cerasis.com/2013/08/08/3pl-vs-4p

*****

Improving HSE in Construction Area by Using Virtual Reality (VR), a Comprehensive Review

Ashkan Karami Mofrad

UPM University, Serdang, Malaysia

Academic Editor: *Amin Teyfouri*

ABSTRACT: The current approach to safety focuses on prescribing and enforcing "defences;" that is, physical and procedural barriers that reduce the workers' exposure to hazards. Under this perspective, accidents occur because the prescribed defences are violated due to lack of safety knowledge and/or commitment. This paper proposes a conceptual, but practical, model of accident causation for the construction industry, highlighting the underlying and complex interaction of factors in the causation process. Accidents occur in all types of construction activities. The accident causation process is complex. Accident prevention requires a comprehensive understanding of this complex process

Keywords: Digital models; Construction safety; Safety practice Design; Risk identification

1.0 INTRODUCTION

This perspective has a limited view of accident causality, as it ignores the work system factors and their interactions that generate the hazardous situations and shape the work behaviors [1]. Therefore, using the appropriate software is necessary to design buildings in order to decrease the casualties in this field [2]. Construction work involves a large number of work processes that need to adapt to the project-specific requirements and context. In contrast to the well-defined procedures of the high-risk systems, the loosely defined construction work processes allow the work crews many degrees of freedom in how they organize and coordinate the work. As a result, construction crew practices determine largely how the actual work is structured and coordinated such as task allocation, sequencing, workload and pace, work coordination, collaborative behavior, etc. and consequently they shape the evolving work situations that the workers face [3].

Concerning the risk and incidents happening unexpectedly at working sites all around the world, construction industry has one of the first ranks between others according to relevant international and official statistics and formal evidences [4].

In most countries, the incidence rate for fatal accidents in the construction industry is higher than any other industry. Snashall [5] found that, on average, five construction workers were killed every 2 weeks and one member of the public was killed every month by construction activities in the United Kingdom. Across the European Union, 67% of workers in the construction sector believe that they are at risk of having accidents [Commission of European Communities (CEC) 1992] [6].It is a great challenge for all those involved in the construction industry to improve this situation by taking effective action to minimize the risk of accidents and ill-health.

2.0 PROBLEM STATEMENT

The occurrence of accidents and injuries continues to be a major problem in construction worldwide. In the past two decades more than 26,000 U.S. construction workers have died at work. That equates to approximately five construction worker deaths every working day. Of these fatalities, 40% involved incidents related to falls from height. Inadequate, removed, or inappropriate use of fall protection equipment contributed to more than 30% of the falls [7].

3.0 OBJECTIVE

This paper decomposed with the aim of fostering and directing further research on the application of digital technologies in design for construction safety, this review article brings together these two strands. After outlining the legislation and practice of design for safety, it first surveys the state-of-the-art in the application of digital technologies – online databases, VR, GIS, 4D CAD, BIM, and sensing/warning technologies – to construction safety in general and to safety through design in particular, articulating the applications that have been developed for use at project, product, process and operation levels [8].

Furthermore, this review looked over on hazard identification and construction H&S competence and assessment in order to indicate the importance of using new technology for preserving HSE in construction area .Because for solving any problem, knowing the causes and how much they can affect one specific issue is first step.

4.0 HAZARD IDENTIFICATION ON CONSTRUCTION PROJECTS

Fig. 1 shows a modified statistical triangle of accident causation. When a hazard occurs it enters the triangle at its base and this is termed a hazardous event. The interior area of the triangle represents the hazardous events; "movement" up the triangle is determined by the severity of the event. The lower boundary condition is termed "near-miss," which exists when the hazardous event resulted in no physical harm, i.e., zero severity. The upper boundary condition is a fatal accident, at the apex of the triangle, which exists when a human life is lost as a result of the hazardous event. An intermediate condition is that of an accident, which exists when the hazardous event does result in physical harm, i.e., severity is greater than zero.

There are two aspects to the control and management of construction hazards: first, the prevention of hazardous events and, second, limiting the potential severity of hazards if they do occur. In terms of Fig. 1, the first of these are preventative control measures that are designed to limit the entry of a hazard into the triangle by reducing its probability of occurrence. The second type is the precautionary control measure, which is designed to limit the movement of the hazardous event within the triangle. This second type reduces risk by reducing the severity of the hazard if it occurs.

Consideration of hazards in terms of their probability of occurrence and severity of consequence provides the general rationale for performing all safety risk assessments, which are undertaken as follows:

1. Estimate the probability of a hazard's occurrence, i.e., its frequency, and its probable severity if it does occur;

2. Evaluate the risk associated with the hazard based upon the frequency and severity estimations; and

3. Respond to the hazard by implementing suitable control measures.

The accident causation and risk control scenario discussed above has one major assumption: that the hazard is identified in the first place. If a hazard is not identified it will have:

1. Complete freedom of entry into the triangle, i.e., the hazard will have an uncontrolled probability of occurrence.

2. Complete freedom of movement within the triangle, i.e., the hazard will have an uncontrolled severity if it does occur [9].

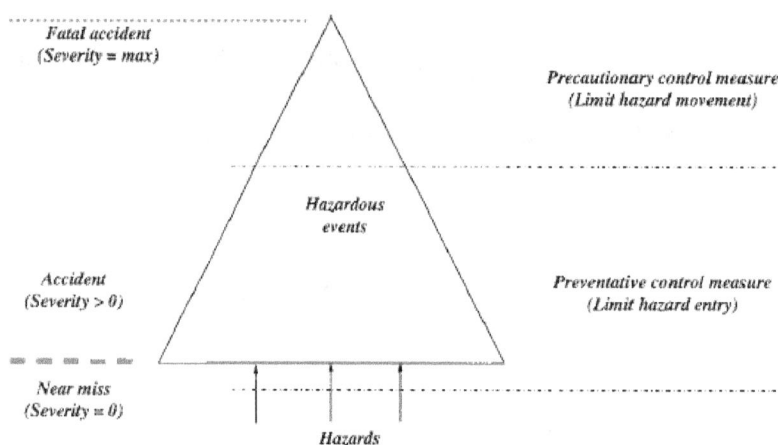

Figure 1: Modified statistical triangle of accident causation (from, Gregory Carter and Simon D. Smith, Safety Hazard Identification on Construction Projects, ASCE, 2006.132:197-205.)

5.0 CONSTRUCTION H&S COMPETENCE AND ASSESSMENT

According to Wright et al. (2003), competence is commonly defined as 'the ability to perform the activities within an occupation or function to the standards expected in employment'. In the context of construction H&S, competence is a two-faceted concept. On the individual level, a competent person is often regarded to be one that has sufficient training, knowledge and experience of the work related. Carpenter (2006a, 2006b) advocates that an individual's H&S competency is the combination of:

Task knowledge (technical or managerial): Appropriate for the tasks to be undertaken.

Health and safety knowledge: sufficient to perform the task safely, by identifying hazard and evaluating the risk in order to protect self and others, and to appreciate general background.

Experience and ability: sufficient to perform the task (including where appropriate an appreciation of constructability), to recognize personal limitations, task-related faults and errors and to identify appropriate actions.

However, the level and development of personal competence within an organization are determined by the organizational H&S culture. Organisations with a real commitment to improving their H&S culture will go beyond ensuring all members of the workforce have received training and are suitably qualified and experienced in the safety-related requirements of their job. A positive H&S culture can enable an organisation to have clear commitments of H&S management, the promotion of H&S standards, effective communication within the organisation, adequate cooperation from and with the workforce and an effective and developing training programme. Thus, an organisation with a positive H&S culture can be seen as H&S competent to undertake a project. Therefore, H&S competence on the organisational level can be defined as 'a culture within an organisation that actively considers the health, safety and welfare of its own people, and of those that its work activities affect, with this being achieved through active management and participation of employee'[10].

6.0 CAUSES OF CONSTRUCTION FATAL ACCIDENTS

Before using any software, we should completely understand the causes of accidents and the amount of their effects. Table 1 is one of the most complete Table which has been designed for this purpose [11].

Table 1: Root Causes of Construction Accidents (from Charlotte Brace Alistair Gibb, health and safety in the construction industry: Underlying causes of construction fatal accidents)

Root Cause	Description	Fatalities (%)
Lack of proper training	An employee was not properly trained in recognizing and avoiding all potential hazards associated with the task he or she is performing	16.6
Deficient enforcement of safety	An employee's supervisor knew that prescribed methods for avoiding hazards were not being followed, but neglected to enforce safety standards	6.8

Safe equipment not provided	An employer does not provide an employee with equipment necessary to minimize hazards	**42.2**
Unsafe methods or sequencing	The normal sequencing of construction tasks does not occur, resulting in a task being inherently more hazardous than is typical	**27**
Unsafe site conditions	Unsafe site conditions The site is inherently more hazardous than are typical construction sites	**0.3**
Not using provided safety equipment	An employee is provided with proper safety equipment, but does not use it properly or does not use it at all	**2.7**
Poor attitude toward safety	An employee may have been properly trained, but does not properly avoid job hazards due to a "tough-guy" mentality, laziness, or a perception that prescribed progress methods would unduly slow job	**0.7**
Solated, sudden deviation from prescribed behaviour	A normally competent and safety-conscious employee suddenly and unforeseeably performs an unsafe act due to fatigue, preoccupation, or	**3.7**

| | likewise | |

7.0 DIGITAL TOOLS FOR MANAGING SAFETY THROUGH CONSTRUCTION

Researchers have developed a range of new tools for use in the construction phase to help contractors achieve safety in their projects. Digital technologies, such as the online databases, VR, GIS, 4D CAD, BIM, sensing/warning technologies etc., are widely applied in these tools for site hazard prevention and safe project delivery. Most of these technologies are combined with each other in related investigations. This section reviews work and highlights the main characteristics of these tools.

The field of Virtual Reality has developed out of experience gained in computer graphics and human-computer interaction. The basic concept is that of the immersion of the operator in a computer-generated world such that the sensory information conveyed reinforces the impression that the model or synthetic world is real. The virtual world is generated by a specialised computing platform and is conveyed to the operator using the human sensory input channels. At the same time, extraneous impressions of the real world are excluded. VR tools for simulating construction processes and site environments fall into the two categories of immersive and non-immersive VR systems [12].

Table 2: Potential Applications of VR in Construction (from, C. Bridgewatera, M. Griffin' and A. Retik' Use of Virtual Reality in Scheduling and Design of Construction Projects, 1994 Elsevier Science)

Area	Applications
	Practicing erection sequences
	Planning lifting operations

Site Operations	Progress and monitoring
	Communications
	Inspection and maintenance
	Safety training and skills
Office Automation	Tele-conferences
	Project review and evaluation
	Project documentation
	Marketing
Design Phases	Initial and detailed design
	Lighting and ventilation simulations
	Data exchange
	Fire/safety/access assessments
	Scheduling and progress reviews
Special Areas	Nuclear industry
	Subsea inspections and work
	Near space operations
	Micro inspection and testing

8.0 IMPLICATIONS OF CURRENT DIGITAL APPLICATIONS FOR CONSTRUCTION SAFETY

This literature has developed a strong set of applications for digital technologies to support construction safety at project, product, process, and operation level. These technologies, as shown in Table 3, and discussed in the previous sub-section(Digital tools for managing safety through construction), include online databases, VR, GIS, 4D CAD, BIM and sensing/warning technologies. Such technologies are often used in combination with each other. Another category is those designed to visualize the construction process to highlight safety implications. The different technologies reviewed are complementary. For example, GIS, on the one hand, provides an overview of construction projects and their associations with their surrounding environment, and is helpful where they encounter hazards from external conditions; BIM, on the other hand, can be used to obtain reusable information from the design phase to clarify construction safety concerns. The combined use of GIS with BIM enables a broader consideration of construction safety.

The researchers found out some of these tools are not without limitations. Take for example, 4D CAD tools indicate a technological approach to planning for safety in advance of operations. A significant shortcoming of this approach is the dependence on computerized construction schedules. Construction operations are dynamic and subject to frequent changes that do not comply with originally scheduled work. Hence, digital schedules are rarely updated sufficiently frequently to accurately reflect all operations underway at any given point in time; though tools that are based on an operational production control model rather than a 4D based process model may reflect this more accurately. An alternative approach is to obtain the most reliable plan possible in the multidisciplinary construction field using collaborative 4D construction planning [13].

Table 3: Set of applications for digital technologies to support construction safety at project, product, process, and operation level (from Wei Zhou, Jennifer Whyte, Rafael Sacks, Construction safety and digital design: A review, 2011 Elsevier)

ol/project	Approach	Level	Technology
&S competence assessment	Assessment of duty-holders competence	Project	Online databases
nstruction Safety and Health Monitoring System (CSHM)	Monitor project performance	Project	Online databases
sign for Safety Process (DFSP)	Simulation and review of construction process for design related safety issues	Process and Product	VR
rtual Construction Laboratory (VCL)	Simulation and review of innovative processes	Process	VR
BA-black building	Safety planning considering environmental conditions	Process	GIS, entity-based 4D CAD
cision Support System (DSS)	Assist monitoring and control of operations	Process	GIS
tterns Execution and Critical Analysis of Site Space Organization (PECASO)	Critical space-time analysis	Process	Entity-based 4D CAD
ıle-based 4D system	Rule-based	Process	Entity-based 4D CAD
antylinna building	Visualization	Process	BIM-based 4D CAD
fety Analysis of Building in Construction (SABIC)	Structural analysis	Process	BIM-based 4D CAD
nstruction Hazard Assessment with Spatial and Temporal Exposure (CHASTE)	Construction job safety analysis and evaluation of operational risk levels	Operation	Entity-based 4D CAD
mputer image generation for job simulation (CIGJS)	Simulation for job safety analysis	Operation	VR
ıtomated obstacle avoidance support system	Sparse point cloud	Operation	Laser range scanning technology
al-time proximity and alert system	Generate active warning or feedback in real time	Operation	Wireless & RFID communication
iFi-based indoor positioning system	Indoor positioning	Operation	Wireless & RFID communication
deo rate range imaging system	Detect, model, and track the position of static and moving obstacles	Operation	Video laser range scanning technology

9.0 DISCUSSION

This paper presented various sorts of software for designing building and preparing safety in construction sites. Managers can decrease the construction casualties by applying this software's; furthermore, this software's based on this research has many merits which mentioned in other sections above. This software can give designers comprehensive framework to design any kinds of structure more safely. In addition, there are not any voids between these software's for designing any specific part of civil engineering since they are involving all the aspects of design for any kinds of structures.

10.0 CONCLUSION

Within this research trajectory, there are significant ways that this research could be extended. For example, research that combines the VR-based CIGJS method with CHASTE could begin to take human factors into account in loss-of-control events. VR technology in the meantime can play a key role in educating and training construction workers. Additionally, existing research on sensing and warning technologies shows their powerful effects on real-time site monitoring to locate various resources; and the potential for further applications. Hence, both individually and in combination there is the potential for significant future research to develop this set of technologies to address construction risks and hazards prevention.

REFERENCE

[1]. P. Mitropoulos, G. Cupido, Safety as an emergent property: investigation into the work practices of high-reliability framing crews, Journal of Construction Engineering and Management 135 (5) (2009) 407–415.

[2]. Akhmad Suraji, A. Roy Duff, and Stephen J. Peckitt, DEVELOPMENT OF CAUSAL MODEL OF CONSTRUCTION ACCIDENT CAUSATION, 2001.127:337-344.

[3]. Panagiotis Mitropoulos1 and Gerardo Cupido2 Safety as an Emergent Property: Investigation into the Work Practices of High-Reliability Framing Crews

[4]. Roozbeh Ghaderi , Mohammad Kasirossafar, Construction Safety in Design Process , ASCE 2011

[5]. Snashall, D. (1990). ''Safety and health in the construction industry.'' British Medical J., London, 301(6752), 563–564.

[6] .Commission of the European Communities (CEC). (1992). Safety and health in the construction sector, Her Majesty's Stationery Ofc., Lon-don.

[7]. A.R. Atkinson, R. Westall, The relationship between integrated design and construction and safety on construction projects, Construction Management & Economics 28 (9) (2010) 1007–1017.

[8] .Y. Mori, B.R. Ellingwood, Reliability-based service-life assessment of aging concrete structures, Structural Engineering 119 (5) (1993) 1600–1621.

[9] .Gregory Carter and Simon D. Smith, Safety Hazard Identification on Construction Projects, ASCE, 2006.132:197-205.

[10]. Wright, M., D. Turner, et al. (2003). Competence Assessment for the Hazardous Industries. Research Report 086. Norwich, Health and Safety Executive.

[11]. Hao Yu M.Sc. B.Eng. , A KNOWLEDGE BASED SYSTEM FOR CONSTRUCTION HEALTH AND SAFETY COMPETENCE ASSESSMENT , 2009

[12]. C. Bridgewatera, M. Griffin' and A. Retik' Use of Virtual Reality in Scheduling and Design of Construction Projects, 1994 Elsevier Science.

[13]. Wei Zhou, Jennifer Whyte , Rafael Sacks , Construction safety and digital design: A review, 2011 Elsevier

A Review on Micro-Manufacturing, Deforming Processes and Their Prominent Issues

Amin Teyfouri

Department of mechanical and manufacturing engineering, Faculty of engineering, UPM, Malaysia

Academic Editor: *Shahryar Sorooshian*

ABSTRACT: Nowadays, there are increasing demand on micro-systems, micro products/components and micro-devices. Thus, the role of micro manufacturing in all aspects of modern industry is taken for granted. Then, typical processes in micro manufacturing are discussed. **Furthermore, one of the most widespread micro-manufacturing processes in deforming processes is micro forming that is presented in this paper. Finally, besides continuing effort in developing micro forming, this paper strives to analyze systematically the key methods of stamping, bending, forging and deep drawing processes and their prominent issues.**

Keywords: Micro-manufacturing; Micro-forming; Micro stamping; Micro forging; Micro bending; Micro deep drawing.

INTRODUCTION

Presently, a plurality of industrial developments are made for compact products. Dramatic changes in during two decades in economic and science have influenced how manufacturing is organized and implemented. The trend toward using micro-scale products for biomedical, electronics, aerospace and defense has been soaring throughout the world. As a matter of fact, Micro-manufacturing engineering is a series of relevant activities within the chain of manufacturing micro-products/features, involving design, analysis, materials, processes, tools, machinery, operational management methods and systems, etc. For product in measure of nanometer and micrometer range, it is no longer trivial to describe how these components are measured, created, handled, assembled and controlled. If micro- products are presented by a huge market it is vital to develop production technology, processes and material that have ability for supporting an industrial production; that is, as Masuzawa mentioned at progression of technology in micro scale, industry has to consider on modern technology in production chain of micro product [1] figure 1. Undoubtedly, a common micro-manufacturing structure is micro forming process; that is, it is widely used in deforming processes for the mass-manufacture of micro-products.

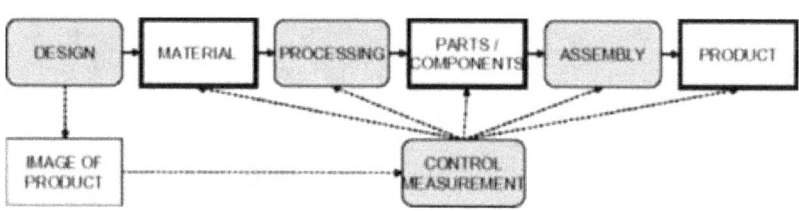

Figure 1: association between objects in the process chain of production and technologies [1]

The forming of miniature-parts or small metal-parts has been evaluated for long time but when sizes decrease to hundreds of microns, precision requirements for small-parts reduce to less than a few microns. Thus, forming of micro scale is required to day-to-day investigation. This paper provides an overview of the approaches perform for the micro-forming method.

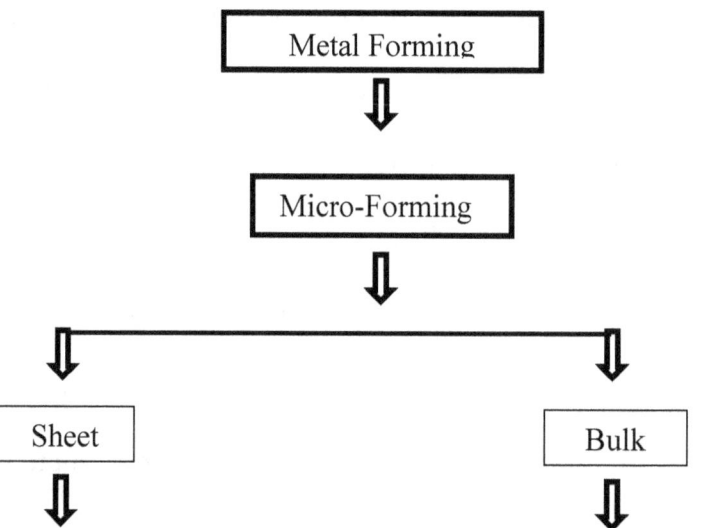

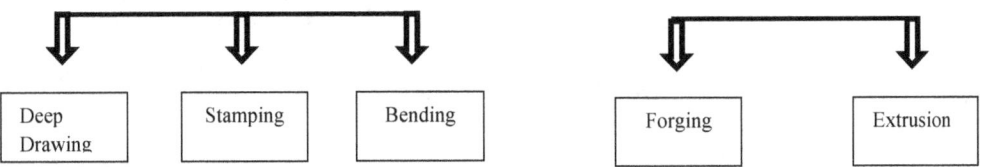

Figure 2: The approaches of micro-forming method

MICRO MANUFACTURING STRUCTURE

MICRO-MANUFACTURING SYSTEMS/METHODS

Obviously, conventional and non-conventional approaches have been widely utilized to manufacture micro-products by miniaturizing or downscaling [2]. In fact, general manufacturing processes can be categorized to subtractive, joining, hybrid and deforming processes in table 1 [3]. The most substantial advantage of micro-manufacturing is the capability for producing parts which having feature sizes of less than 100μm [4] [5] [6] or little larger than the thickness of a human hair.

Table1: processes in micro-manufacturing [3]

Subtractive processes	Micro-Mechanical Cutting (polishing, grinding, turning, milling, etc.), Photo-chemical-machining, Laser Beam Machining, Electro Beam Machining.
Joining processes	Micro-Mechanical-Assembly, Gluing, Resistance, Bonding, Laser-welding, Vacuum Soldering, laser, etc.

Hybrid processes	Micro-Laser-ECM, Micro-EDM and Micro-ECM and Laser assembly, Combined micro-machining and casting, Laser-assisted-micro-forming, Shape Deposition and Laser machining, LIGA and LIGA combined with Laser-machining, Micro assembly injection moulding, etc
Deforming processes	Micro-forming (stamping, bending, deep drawing, forging, extrusion, hydro-forming, incremental forming, superplastic forming, etc. Hotembossing, Micro/Nano-imprinting, etc.

Micro-mechanical cutting is a fabrication procedure for producing miniature devices and components with feature of tens of micrometers to a little millimeters in size. The motivation for micro-mechanical cutting stem from the translation of macro-process knowledge to the micro-domain [7]. Precision cutting tools are so important due to feature and quality of micro-structures. The majority of micro-machine tools are created from ultra-precision material as well as high rigidity. Even though this approach have been encountered with some challenging problems such as producibility, predictability, and productivity

[8]. Recent researches demonstrate use of cutting force in micro-machining can help to improve and monitoring the quality of sculptured products [9] in addition industrialized analytical cutting force model aid operators to select the right cutting condition for their organization or system. Furthermore, the elastic-plastic deformation of the work piece also changes the cutting force in micro-machine operations. Typical micro-cutting operations include micro-milling, micro-turning, micro-drilling and micro-grinding.

Micro-mechanical assembly is an approach for assembling components on micro-scales, where the relative location of components is kept by exchange of contact forces providing by mechanical limitation. Main processes for micro-assembly and packaging involve micro-casting/moulding, micro-welding, mechanical placement/insertion/ pressing, resistance/laser/vacuum soldering, bonding, gluing, etc. Interconnection and packaging solutions are a vital technology for connecting micro-systems to the macro world [10]. The base of mechanical assembly is classified to three purposes of handling and positioning with the function of setting two or more items into specific mutual position and orientation, mechanical assembly with the goal of ensuring the mutual connection between components against outside properties and quality control for ascertaining whether the mechanical assembly process has been carried out as specified [11]. Obviously, the both general domain of automated and manually assembly of micro devices include handling of parts that are extremely small (in the order of 10^{-6} m). Moreover, they take into account increase of reliability, efficiency as well as reducing cost [12]. Even though substantial progress has been created in the manufacture of individual micro components/parts, there is still a major amount of manual work included in assembly/packaging of the micro-products and systems. There are prominent issues in micro-assembly such as lack of the guidelines and the requirements for tailor-made tools for the processes.

Micro-machining technology has an exclusive opportunity to produce micro-components that are so flexible and cost-effective and it is noted that recognition correct method in machining has significant role in preventing defect specially in micro scale; i.e., some types of method such as contact measurement is not suitable for micro fabrication since the structure of micro parts are low rigidities however this method is the most certain method to change the shape of work piece. Incontrovertibly, optical technique is appropriate for quantitative measurement [13]; for instance, **accenting laser micro machining of up to 1 mm thick metal should consider on advanced performance.** In laser's characteristics such as pressure, the movement of optics and piece, laser beam alignment, temperature, energy distribution, combination of pulse with wavelength for penetration depth and providing optical and other characteristics must be optimized for development of a laser processing scheme in micro-machining [14].Moreover, Electrical machining is special method in micro technology such as **micro-electrical discharge machining (EDM)** and **micro**-electrochemical machining (**ECM**). In ECM and EDM, the main goal is to forecast the shape of the workpiece. EDM with the famous title of laser drilling is basically a thermal process and consequently thermal modeling is so imperative while ECM computes the work dissolution rate with function of **determination of the current density distribution** [15].

MICRO FORMING METHOD

Micro-products often are introduced by process of plastic forming technology [16].Various process in micro- forming method is included stamping, forging, hydro-extrusion, extrusion, bending, superplastic forming, deep drawing, etc. These processes is implemented by real scaling down of the conventional structures, tools and machines with additional maintenances [17]. Micro-forming can be one of the best options in the mass-manufacture of micro-parts, where the micro-forming machine is autonomous and it is integrated with parts transport systems and high precision feeding (Akhtar Razul Razali, A Review on Micro-manufacturing, Micro-forming and their Key, 2013). In fact, quality and efficiency are two important part in handling devices, machines and forming tools in industrial application of micro-forming. Heating has been widely used for micro-forming process since it helps to higher strength or extend forming ability in terms of part and dimensions and it is really common in micro-forming process. Main problems has to be addressed in this method include qualification of forming limits, tool/material interfacial conditions, process design optimization, process modeling and analysis, understanding of material deformation mechanisms and material property characterization [10].

The figure 4 shows some achievements are come from some types of approaches in micro-forming method.

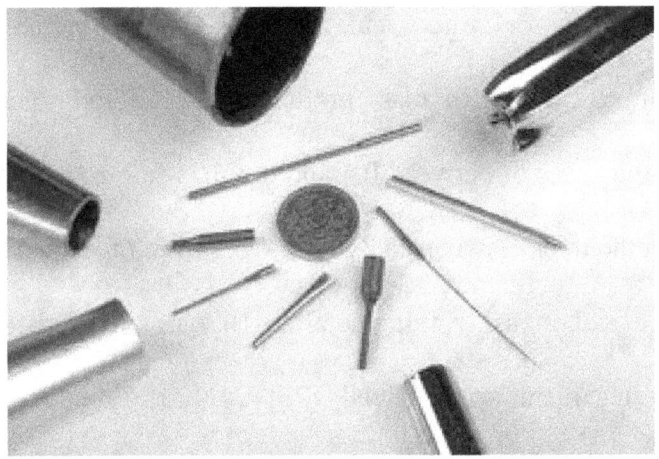

Figure 4: common achievement has been produced by micro-forming

THE APPROACHES OF MICRO-FORMING

1.0   STAMPING AND SHEET METAL STRUCTURE

It is noted that stamping is a prominent method in forming process. Nowadays, Metal stamping manufacturing processes are popular because they are an economical and quick means of producing intricate, accurate, strong and durable articles in huge quantities [19]. In fact, metal stampings are used in most of the mass-produced product with different methods. According to a survey in the US, every

American home has almost 100 000 metal stamping process in the 1980s [20]. Metal stamping has been entered in the manufacturing of metal components with a particular configuration from sheet metal stock and a various range of processes such as embossing, punching, flanging, blanking, coining, and bending [21]. Stamping can be done on metals such as titanium, bronze, aluminium, zinc, nickel, steel, Inconel, copper, and other alloys (Akhtar Razul Razali, A Review on Micro-manufacturing, Micro-forming and their Key, 2013). Sheet metal components are used numerous applications such as aircraft, electronics products and for packaging such as consuming goods and computer screen etc. Sheet metal parts include electrical connectors and lead frames, micro-meshes for masks and optical devices, micro-springs for micro-switches, micro-cups for electron guns and micro-packaging, micro-laminates for micro-motor and fluidic devices, micro-gears for micro-mechanical devices, casings/housings for micro-device assembly/packaging, micro-knives for surgery, etc. Sheet metal, stamping are one of the most effective approaches to meet the future demand of mass production [22]. Traditionally, sheet metals may be defined as metal having a thickness of between 0.4 and 6 mm, while micro-sheet forming usually deals with sheet metals of which the thickness is usually below 0.3 mm. Previous academic results show that stamping processes can be real in using thin metal in Bipolar plates that are one of the essential components of proton exchange membrane fuel cells [23] [24] [23].

Micro stamping

It may indeed be true to say that micro-stamping is a target for producing miniaturized components and products. Micro-stamping is valuable for producing some components such as wristwatch and micro handheld-device components, medical products and so on [21]. Development of micro stamping in process have shown the effort for performing an automatic hybrid and simple punching operation on brass strips and continued movements in this development influence on manually-operated micro-stamping machine with the capability of employing various punch shapes and progression of micro stamping demonstrates the widespread use of this process in micro-sheet-forming [25].

A micro-stamping test machine can be useful methods for progressive stamping chain, for example, dry micro-stamping machine designed to be hard enough in order to operate in down-sized configuration ,moreover , it has characterized to have maximum loading capacity up to 5 KN as well as maximum 100rpm speed [26]. As can be seen in figure 5, progressive stamping chain in this test is collected of shearing, ironing and bending operations. Actually, the holes were punched out without substantial burrs in shearing process and the stage of ironing and bending might happen severe wear between worksheet and punch in ironed surface. The surface was evaluated by fine micrometer and optical microscope [26].

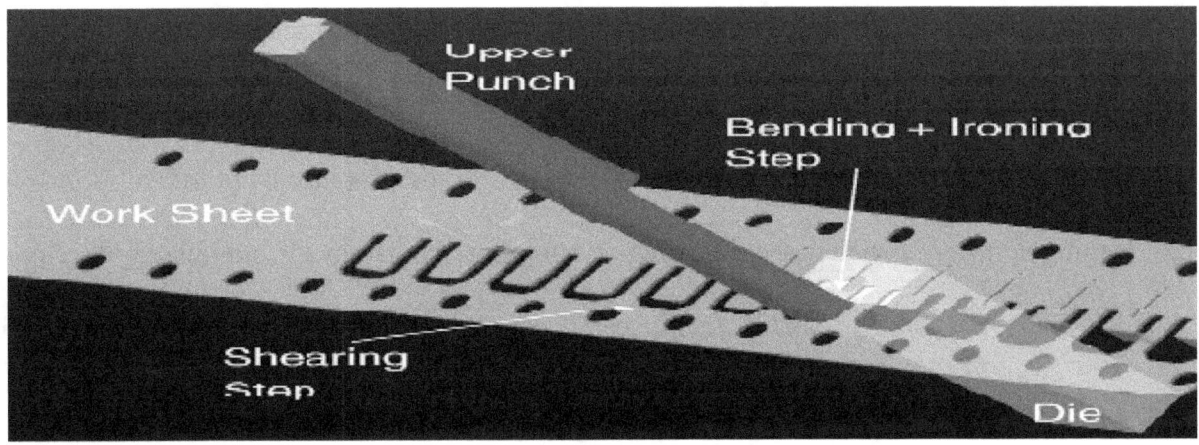

Figure 5: A micro-stamping test in the progressive stamping sequence [26]

Generally speaking, improvement of micro stamping needs to efforts to realize operation in terms of it process, tooling, and particularly material-handling. In addition, attention to micro material's sizes is a key point of stamping structure for modern manufacturing technology.

2.0   FORGING STRUCTURE

Nowadays, forging methods are planned to being performed systematically in controlled presses and hammers to produce forged shapes with a high degree of dimensional accuracy and structural integrity. Statistics published by the Forging Industry Association show that forgings are used in 20% of all products that represent the U.S. Gross Domestic Product, and that the forging industry has annual sales of US$ 6 billion and employs 45,000 personnel in North America alone [27]. The forging industry is keeping pace as with other metal forming processes through continuous progress in many areas. Actually, six characteristics play important roles in forging process that include the development of alloys in processing or forgeability, growing the industry's understanding of the mechanics of the forging process as production rates ought to be increased, and costs reduced. The third characteristic is some forging companies have been using systems for controlling critical processes as reducing variables, improving dimensional precision, and eliminating costly chip-making operation. Using CAD\CAM throughout production processes and design to developing dimensional accuracy of forgings, forging companies should use modeling and forging simulation for minimizing development time. Finally, fast tool change capability facilitates the preplanned replacement of die inserts in long production runs, and reduces changeover time for shorter runs [28].

Forging process

Forging is probably oldest metal working process and was known even during prehistoric days when metallic tools were made by heating and hammering. Forging is basically involves plastic deformation of material between two dies to achieve desired configuration and depending upon complexity of the part forging is carried [29]. Its noted that this process is particularly problematic as the high costs and long lead times is associated with die tooling represent serious obstacles to forging competitiveness [30] and another problem is the limit it places on product and design flexibility [31]. As a matter of fact, the classification of modern forging process includes open die, impression die, ring rolling, warm forging and cold forging [28]

- Open die forging

Open die forging is hot forming process that usually worked for producing large parts. The shapes of open die forging usually are V-shape, semi-rounds, or flat. There is basically no limit to the size of a forging made with the open-die method. Nonetheless, most work will need extensive machining to achieve their shape or net shape. During the process of open die forging due to heavy duty and mechanical material handling, normally it is required to cranes, fork lifts, and rotating devices [28].

- Impression die forging

Utilizes a pair of matched dies with contoured impressions in each die. When the dies close, the impressions form a cavity in the shape of the forging. Often two or more progressive impressions are used, sometimes in conjunction with one or more preforming operations, to form the desired shape, in other words, the workpiece is deformed between two die halves which carry the impression of the desired final shape and it usually uses for smaller components. The suitable forging temperature improves plastic flow characteristics and reduces the forces on the forging tools [28].

- Ring rolling forging

Ring rolling first scientific developments were made in the 20th century [32]. The process begins with a "donut" shaped preform, which is made by upsetting and piercing operations. The preform is placed over the idler roll in a ring rolling mill. The idler roll is moved toward a drive roll, which rotates to reduce the wall and increase the diameter, while forming the desired shape [28]. Long process times and high process cost are disadvantages of these methods [33] while it is perfect method for microstructure by continuous local plastic deformation [34]. Ring rolling process which allows a flexible near-net-shape forming of both hot and cold rings is presented by Allwood et al [35].

- Cold forging

Cold forging has numerous influence on manufacturing process in huge quantities, parts with surface quality and high dimensional accuracy in addition to high strength because of work hardening. Material costs and machining resources can be saved by this type of net-shape production [36].

- Warm forging

Warm forging is a modification of the cold forging process where the workpiece is heated to a temperature significantly below the typical hot forging temperature [28].

3.0 BENDING STRUCTURE

Bending in sheet metal forming explain as the straining of the metal around a straight axis. A neutral axis plane exists for the sheet metal around which the top section of the material may be stretched during bending while the lowest part become smaller by compressing. Usually in conventional sheet metal working, bending operation is depending on bending process that may use wipe dies, punches, rolls, the downward movement of the bending tools for creating edge bending, U-bending, V-bending, etc.

Micro-Bending Process

Microbending (bends too small to be seen with the naked eye) happen when pressure is applied to the surface of an optical fiber [37]. The use of bending process in manufacture of micro sheet metal products mainly is for electronics products and MEMS (micro electromechanical manufacturing systems), as industrialization of regular requirements such as producing 3D profiles/sections [38]. Totally, for creating micro products can mention to the key parameters of the bend angle, bend radius, bend allowance, length of bend, etc., in some stages of bending process.

Controvertibly, the plan and design of typical micro sheet metal forming process specially for bending procedure must be done size-dependent material behavior to fulfil accuracy requirements. One of the common problems in bending of micro-scale sheet metal is to stop the changed appearance of the sheet and controlling springback-related issues, in the figure 1. This situation often happens when micro-components with shortened structure or bending/cutting process are produced while they still have a progressive die-stamping/forming pattern. The parts from errors caused by the springback may replace for correct designing the die and the bending characteristics, or introducing additional process. Due to the fact that the parameters of micro-parts, micro-assembly is hard to handle, therefore making spring back or elimination of them is more crucial than in the macro scale. Thus, bending test helps to show the result of these size effects on springback behavior of metal [39]. Stretch bending, overbending, bottoming and rebending are the methods usually utilized to remove the errors resulted from the springback in conventional sheet metal forming. Another indispensable factor can influence on reducing springback is the changing of product design and original structure of production because it allows to obviate occurrence of springback before the step of production.

It may be difficult to use some of these various methods in micro-sheet forming, for instance difficulty in adding additional tools, limited cost, difficulty of access to the geometric small areas, forming parts in the limited tooling space and forming special material. As a case in points, the behavior of bending process in metal foil are basically affect by two opposing size reason [40]: First is a reduction in material strength result from the increasing share of surface grains on the overall volume [41]. Next reason is the declining foil density at the most continual permissible strain led to the increasing thickness of geometric-dependent dislocation because of larger strain gradients [42],for avoiding difficulties in handling parts/microcomponents and in the fabrication of microtooling, non-contact processing methods such as laser-assisted bending may be suggested to achieve accurate bend geometries [43];for example, in several years ago Vollertsen and Kals investigated the micro sheet metal forming in technical aspect of laser bending and air bending for thickness of down 0.1mm [44] [45].

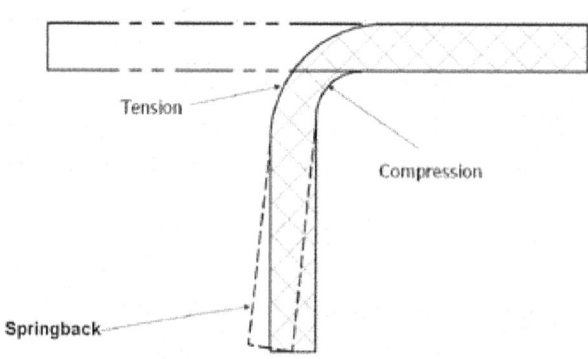

Figure 6: Phenomenon of springback and bending process [46]

4.0 DRAWING STRUCTURE

Drawing is one of the main sheet metal-forming processes that utilize to make totally complex-curved, hollow-shaped sheet parts, or box-shaped, cup-shaped [47]. As a matter of fact, drawing sheet metal is a part of metal science for setting correct balance between breaking to reach a successful part and wrinkles. Drawing process totally is divided to the two types: first is sheet metal drawing that includes plastic deformation in curved axis and second is tube, bar, wire drawing that involves to decline in diameter of a die and growth of its length. Actually, impressive drawing during the process depends on a number of reasons, these involve the result of the changing blank holder force to stop wrinkles without excessively delaying metal flow, the choose of suitable material for both blank and die to have adequate coefficient of friction, formability of the material being drawn, limited punch force in drawing operation to lower worth than that which will fracture the shell wall [48].The information about the geometric change and stiffness attributes is vital due to the purpose of the process limits of the deep drawing. Furthermore, this knowledge is necessary for investigation of stiffness decrease for real components in initial structured sheet metal [49].

Micro-deep drawing

The process of micro deep drawing are explained to a sheet that place on a die as well as sufficient amount of pressure apply to a blank holder. With the move downward of the punch, the sheet push into the die and the areas formed after drawing [50]. The first approach of micro deep drawing parts for deeper understanding according to the theory of similarity is downscaling the factors of mechanical macro deep drawing [51].Normally, deep drawing process may combine some operations such as compression, shearing, bending, unbending and stretching that is associated with geometric part to be manufactured. Thus deep drawing is a more complex process than bending or cutting/shearing process, but the process of deep drawing is quite similar to cutting operation.

Drawing ratio, diameter of the punch/ the blank = DR, is a major problem of micro-deep drawing due to DR values in limited space, which normally creates limitation of tooling arrangement and controlling conditions. This limiting drawing ratio (LDR) generally is associated with micro-structure and sheet material thickness and also the method of changing blanks and formed cups with the sheet metal strips in layout design of stamping/forming; in fact, it is performed because of the difficulties are related to handling these small item. Experimental investigation of micro deep drawing demonstrate the basic issue of limit drawing ratio that stem from problems of size-effects; in other words, this situation often occur when friction coefficient in micro deep drawing that resulted from applied pressure. In micro deep drawing process due to size-effect and its new concerns, it is compared to macro deep drawing while experiments have shown under the same forming conditions the friction coefficient in micro deep drawing is greater than that in macro deep drawing [52] [53] [54] [55]. One of the most processes that produce significant friction in micro deep drawing is strip drawing in figure 1, when the punch start to move into female die, between the blank holder and the sheet, and between female die and the sheet, friction occurs [56]. Even though it has numerous friction in process, the total goal of strip drawing can be simply to improve dimensional tolerances, to improve surface, or to work hardly and so rise the strength of the product [57].Recent research have indicated, most of the results in this challenge have been investigated by powerful tool Finite Element (FE) simulation and it is beneficial to forecast the performance of processes in advance, but FE-programs don't consider size effect. Therefore, mathematical model for implementation and size effect is necessary [58].

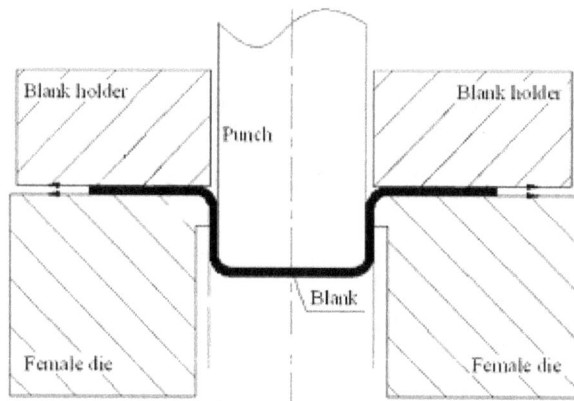

Figure 7: performance of deep drawing and schematic of strip drawing [59]

Redrawing is normally essential for micro products, because of limited accomplishment of a possible reduction value of a cup. Redrawing operation can demonstrate ironing and annealing process is feasible for miniature cups [60].When micro products are encountered with limiting drawing ratio, redrawing process has to be employed for improving this problem; that is, If the first redrawing cannot manufacture a deep cup, second redraw is worked, and so on [61]. Figure 2 indicates a redrawing process. As paper demonstrates the redrawn product is performed on the inner fringe b0a0. At the same time during this process, the material on b"a" at a sure height on the partition of the pre-drawn cup has to reach and pass through the inner fringe b0a0.

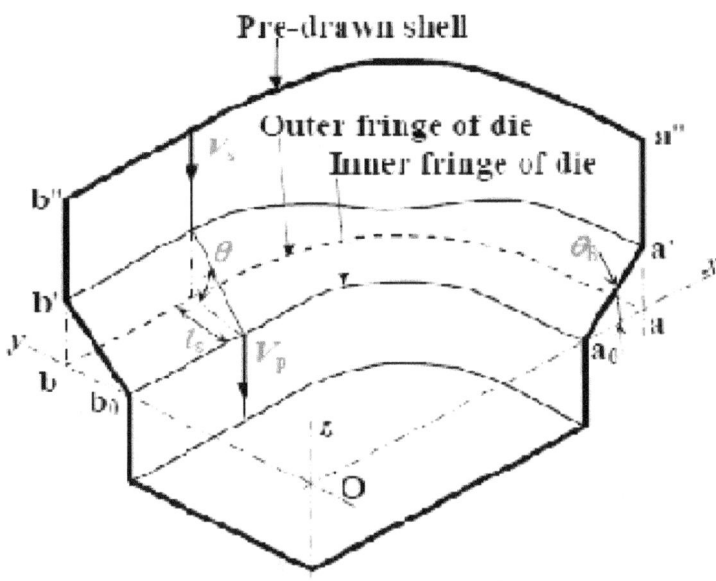

Figure 8: Redrawing of rectangular cup [62]

CONCLUSION

Incontrovertibly, the researches of micro-manufacturing have been produced substantial results in processes, material and equipment during 2000 up to now. In micro-components, the high-volume production should be prominent purpose for designing micro-manufacturing. As discussed in this paper, further analysis is necessary for improving micro-forming method particularly deep drawing and bending process and also more fundamental researches are required on interaction between handling tools and materials. In future, micro-manufacturing will be better connection between nano-manufacturing and macro-manufacturing due to progression of technology in machining and handling.

REFERENCES

[1] T.Masuzawa, "State of the Art of Micromachining," *Manufacturing Technology,* vol. 49, no. 2, pp. 473-488, 2000.

[2] M. G. ,. A. P. ,. E. C. a. C. G. Aldo Attanasio, "Influence of Material Microstructures in Micromilling of Ti6Al4V Alloy," *materials,* pp. 4268-4283, 2013.

[3] Y. Qin, "Overview of manufacturing," in *Micromanufacturing Engineering and Technology,* Elsevier, 2010.

[4] S.-H. R. S.-I. O. Byung-Yun Joo, "Micro-hole fabrication by mechanical punching process," *Materials Processing Technology,* vol. 170, no. 3, p. 593–601, 2005.

[5] J.-C. C. G.-L. Renn, "Development of a Novel Micro-Punching Machine Using Proportional Solenoid," *CHINESE SOCIETY OF MECHANICAL ENGINEERS,* pp. 89-93, 2004.

[6] Y. C. Gwo-Lianq Chern, "Study on vibration-EDM and mass punching of micro-holes," *Materials Processing Technology,* vol. 180, no. 1_3, p. 151–160, 2006.

[7] S. P. T. F. J. Chae, "Investigation of micro-cutting operations," *Machine Tools & Manufacture,* p. 313–332, 2006.

[8] K. C. D. W. F. W. Xichun Luoa, "Design of ultraprecision machine tools with applications to manufacture of miniature and micro components," *Materials processeing technology,* vol. 167, no. 2_3, p. 515–528, 2005.

[9] M. J. R. D. S. K. X Liu, "Cutting mechanisms and their Influence on dynamic forces, vibrations and stability in micro-endmilling," *Mechanical Engineering Congress and Exposition.,* p. 13_20, 2004.

[10] A. B. Y. M. A. R. J. Z. C. H. W. P. X. D. D. L. Yi Qin, "Micro-manufacturing: research, technology outcomes and development issues," *Advanced Manufacturing Systems and Technology ,* p. 821–837, 2010.

[11] M. A. G. T. a. A. G. Hans Nørgaard Hansen, "Micromanufacturing Engineering and Technology," in *Micro-Mechanical-Assembly*, Elsevier, 2010.

[12] D. P. D. V. J. Cecil, "Assembly and manipulation of micro devices—A state of the art survey," *Robotics and Computer-Integrated Manufacturing,* vol. 23, p. 580–588, 2007.

[13] T. Nagayoshi Kasashima, "Laser andelectrochemicalcomplexmachiningofmicro-stentwith on-machine three-dimensionalmeasurement," *Optics andLasersinEngineering,* vol. 50, p. 354–358, 2012.

[14] T. K. N. M. D. A. R. D. H. E. D. Jahns, "Laser trepanning of stainless steel," *Physics Procedia,* vol. 41, p. 630 – 635, 2013.

[15] M. K. S. Hinduja, "Modelling of ECM and EDM processes," *CIRP Annals - Manufacturing Technology,* vol. 62, p. 775–797, 2013.

[16] Y. Qin, "Micro-forming and miniature manufacturing systems — development needs and perspectives," *Materials Processing Technology,* vol. 177, no. 1_3, p. 8–18, 2006.

[17] G. L. A. M. M. S. J. M. X. H. R. R. A. D. K.P. Rajurkar, "Micro and Nano Machining by Electro-Physical and Chemical Processes," *CIRP Annals - Manufacturing Technology,* vol. 55, no. 2, p. 643–666, 2006.

[18] Y. Q. Akhtar Razul Razali, "A Review on Micro-manufacturing, Micro-forming and their Key," *Procedia Engineering,* vol. 53, p. 665 – 672, 2013.

[19] A. N. B.T. Cheok, "Trends and developments in the automation of design and," *Materials Processing Technology,* vol. 75, p. 240_252, 1998.

[20] A. Nee, "PC-based computer aids in sheet-metal working," *mechanical working technology,* vol. 19, no. 1, p. 11–21, 1989.

[21] S. S. R. Kalpakjian S, Manufacturing Engineering and Technology, 2006.

[22] J.-M. Y. Tsung-Chia Chen, "Fabrication of micro-channel arrays on thin stainless steel sheets for proton exchange membrane fuel cells using micro-stamping technology," *Advanced manufactruring technology,* vol. 64, p. 1365–1372, 2013.

[23] M. K. M. H. Toyoaki Matsuura, "Study on metallic bipolar plate for proton exchange membrane fuel cell," *Power Sources,* vol. 161, p. 74–78, 2006.

[24] B. G. D. S. Jie Xu, "Size effects in micro blanking of metal foil with miniaturization," *Advanced manufacturing technology,* vol. 56, no. 5_8, pp. 515-522, 2011.

[25] Y.-J. E. W. S.-F. L. Gwo-Lianq Chern, "Development of a micro-punching machine and study on the influence of vibration machining in micro-EDM," *Materials Processing Technology,* vol. 180, no. 1_3, p. 102–109, 2006.

[26] E. I. K. I. Tatsuhiko Aizawa, "Development of nano-columnar carbon coating for dry micro-stamping," *Surface & Coatings Technology,* vol. 202, p. 1177–1181, 2007.

[27] FIA, "Forging Industry Vision of the Future," *Forging Industry Association,* 1996.

[28] "http://www.forging.org/Design/page1.html," [Online].

[29] D. P. Pandey, "http://paniit.iitd.ac.in/~pmpandey," [Online].

[30] F.H. Osman, "Preform Design for Forging Rotationally Symmetric Parts," *CIRP Annals - Manufacturing Technology,* vol. 44, no. 1, p. 227–230, 1995.

[31] F. K. B. D. B. S. Joseph Domblesky, "Welded preforms for forging," *Materials Processing Technology,* vol. 171, p. 141–149, 2006.

[32] K. ,. Weber, "Ring rolling and the construction of ring mills.," *Stahl und Eisen,* vol. 79, p. 1912–1923, 1959.

[33] N. C. T. W. H. B. Berns H, "Herstellung und eigenschaften thermischer spritzschichten mit gradierter structur," *Härterei-technische Mitteilungen,* vol. 48, p. 20_24, 1993.

[34] Z. Z. L. H. Dongsheng Qian, "An advanced manufacturing method for thick-wall and deep-groove ring—Combined ring rolling," *Materials Processing Technology,* vol. 213, p. 1258– 1267, 2013.

[35] K. R. M. D. M. O. Ö. M. S. T. T. A. Allwood J, "The technical and commercial potential of an incremental," *CIRP Annals - Manufacturing technology,* vol. 54, p. 233–236, 2007.

[36] U. E. M. M. T. Kroiß, "Comprehensive approach for process modeling and optimization in cold forging considering interactions between process, tool and press," *Materials Processing Technology,* vol. 213, p. 1118– 1127, 2013.

[37] V. Watson, Microbending & Macrobending Power Losses in Optical Fibres, www.optronicsnet.com, 2011.

[38] S. H. A. S. H.-W. Jeong, "Microforming of three-dimensional microstructures from thinfilm metallic glass," *Microelectromechanical systems,* vol. 12, p. 42–52, 2003.

[39] C. P. M. Y. Jenn-Terng Gau, "Springback behavior of brass in micro sheet forming," *Materials Processing Technology,* vol. 191, p. 7–10, 2007.

[40] U. E. M. M. M. G. Alexander Diehl, "Size effects in bending processes applied to metal foils," *production Engineering Resource Development,* vol. 4, p. 47–56, 2010.

[41] A. MF, "The deformation of plastically non-homogeneous materials," *Philosopical Magazine,* vol. 21, p. 399–424, 1970.

[42] H. J. Fleck NA, "Strain gradient plasticity," *Advances in Applied Mechanics ,* vol. 33, p. 295–361, 1977.

[43] M. N, "Laser micro-bending for precise micro-fabrication of magnetic disk-drive components," *Laser Precision Microfabrication,* p. 24–29, 2003.

[44] F. V. R. K. M. Geiger, "Fundamentals on the manufacturing of sheet metal microparts," *Advanced Performance Materials,* vol. 3, pp. 123-151, 1996.

[45] R. Kals, "Fundamentals on the Miniaturization of Sheet Metal Working process," *PHD thesis,* 1998.

[46] A. Hedrick, 2007.

[47] R.K.Jain, "production technology," *Khanna publishers,* 2006.

[48] A.Purushotham, "Simulation studies on Deep Drawing Process for the Evaluation of stresses and strain," *Computational Engineering Research,* vol. 03, p. 40_47, 2013.

[49] R. O. B. V. V. M. V. Malikov, "Experimental study of the change of stiffness properties during deep drawing of structured sheet metal," *Materials Processing Technology,* p. 1811– 1817, 2013.

[50] B. G. C. W. D. S. Feng Gong, "Micro deep drawing of micro cups by using DLC film coated blank holders and dies," *Diamond & Related Materials,* p. 196–200, 2011.

[51] O. P. H. Pawelski, Technische Plastomechanik, Verlag Stahleisen,, 2000.

[52] H. S. N. H. W. Z. H. F. Vollertsen, "SHEET METAL MICRO FORMING," in *6th international conference on industrial tools and material processing technologies*, bled, slovenia, 2007.

[53] Z. H. H. S. N. C. T. F. Vollertsen, "State of the art in micro forming and investigations into micro deep drawing," *Materials Processing Technology,* p. 70_79, 2004.

[54] K. Y. H. K. Y. Saotome, "Microdeep drawability of very thin sheet steels," *materials processing technology,* p. 641_647, 2001.

[55] B. Y. W. C. M.W. Fu, "Experimental and simulation studies of micro blanking and deep drawing compound process using copper sheet," *materials processing technology,* p. 101_110, 2013.

[56] B. G. D. S. X. B. Chunju Wang, "Tribological behaviors of DLC film deposited on female die used in strip drawing," *Materials Processing Technology,* p. 323_329, 2013.

[57] I. H. H.L. Costa, "Effects of die surface patterning on lubrication in strip drawing," *materials processing technology,* p. 1175–1180, 2009.

[58] U. E. R. K. F. V. A. Messner, "Size effect in the FE-simulation of micro-forming processes,," *materials processing technology,* p. 371–376, 1994.

[59] B. G. D. S. X. B. Chunju Wang, Tribological behaviors of DLC film deposited on female die used in strip drawing, Materials Processing Technology, 2013.

[60] A. B. J. Z. A. R. Y. M. C. H. Yi Qin, "Forming of Micro-Sheet-micro components," in *micro manufacturing,* 2010, p. chapter 8.

[61] C.-L. L. You-Min Huang, "An elasto-plastic finite-element analysis of the metal sheet redrawing process," *Materials Processing Technology,* p. 331_338, 1999.

[62] K. Isobe, "FLASH DESIGN OF DIE SHAPES FOR REDRAWING RECTANGULAR CUPS," *material forming,* vol. 3, p. 85– 88, 2010.

*****

Robotics Application In Civil and Management Engineering
Nima Bahiraei

UPM University, Serdang, Malaysia

Academic Editor: *Amin Teyfouri*

ABSTRACT: In this paper a number of recent developments in the field of civil and management engineering, mining are reviewed. Robotics application is growing in the manufacturing engineering. But in the field of civil engineering and construction to automation technologies has been slower. The most automation application plants for batching concrete, bending steel, and making pre-cast concrete elements. Summary of a variety of robotics application in civil and management engineering are provided. This review describes on how robots industrial are utilized in the management and civil structure.

Keywords: Robotics application; assembling wall segments; CAD and CAM systems

1.0 INTRODUCTION

According to the Robot Industry of America, robot is a mechanical and electronically that can be programmed to do task under control. On the other hand, the civil engineering covers progresses planning, design, and construction and about management activities, including structures, tunnels, mines and transportation systems. Today, robots are a vital tool for the civil in the most countries, but they can perform to limited inspects of the total process in civil engineering. Against this opinion, Germany and Japan are developing extremely in Robotics Application in civil engineering. A variety of application in civil engineering, construction robotics is described below.

2.0   AUTO – DRIVE AND AUTOWORK MACHINES

Smart and unmanned trucks are one of the most interesting current developments of civil engineering and especially mining. Australia is the first country in the world to begin large driverless vehicle application in mining sector and technology is developing rapidly.

3.0   ROBOTICS AND PROTOTYPES

A number of relevant robotic prototypes have been designed and built in United States for applications, which are potentially transferable to the construction industry. In Table 1 we can see list of some of the robots developed to date. These include areas of underground mining, underwater, inspection and search, nuclear power plant, structural assembly in outer space, military application and other.

Table 1: Examples of Robotics Developed in the United States [1][2][3][4][5][6]

Robot Type	Application	Developer
John Deere Excavator, model 690C	Teleported excavation for rapid airport runaway pair	John Deere, Inc., Moline, Illinois
Laser – aided Grading System	Automatic grading control for high – volume earthwork	Gradeway Construction Co. & Agtec Development Co., San Francisco, California; Spectera -Physics Co., Dayton Ohio
Automatic Slipform Machines	Placement of concrete sidewalks, curbs and gutters	Miller Formless System Co., McHenry, Illinois; Gomaco, Ida Grove, Iowa
Micro – Tunneling Machines	Teleported micro tunneling	American Augers, Wooster, Ohio
Robotic Excavator (REX) and Autonomous Pipe Mapper	Autonomous excavation around buried utility metallic pipes, potentially for several types of autonomous non-destructive testing	Carnegie Mellon University, Pittsburgh, Pennsylvania

''NavLab''	Autonomous navigation in unstructured terrain	Carnegie Mellon University, Pittsburgh, Pennsylvania
Remote Work Vehicle	Nuclear accident recovery work, demolition of structures after nuclear accidents, structural surface decontanamination, clean up and treatement, transport of materials.	Carnegie Mellon University, Pittsburgh, Pennsylvania

4.0 DESIGNS AND CONSTRUCTION OF THE SLIDING DOMES

The Prophet's holy mosque is in Madinah. This project is a successful concept from other industries to civil and management engineering. The first level of the process started the use of CAD (Computer Aided Design). In this process, design tasks such as the finite-element structural analysis and design optimizations were integrated through the CAD tool. The next level involved in the direct utilization of Computer Aided Machining (CAM) in the production process where incredible advantages could be realized.

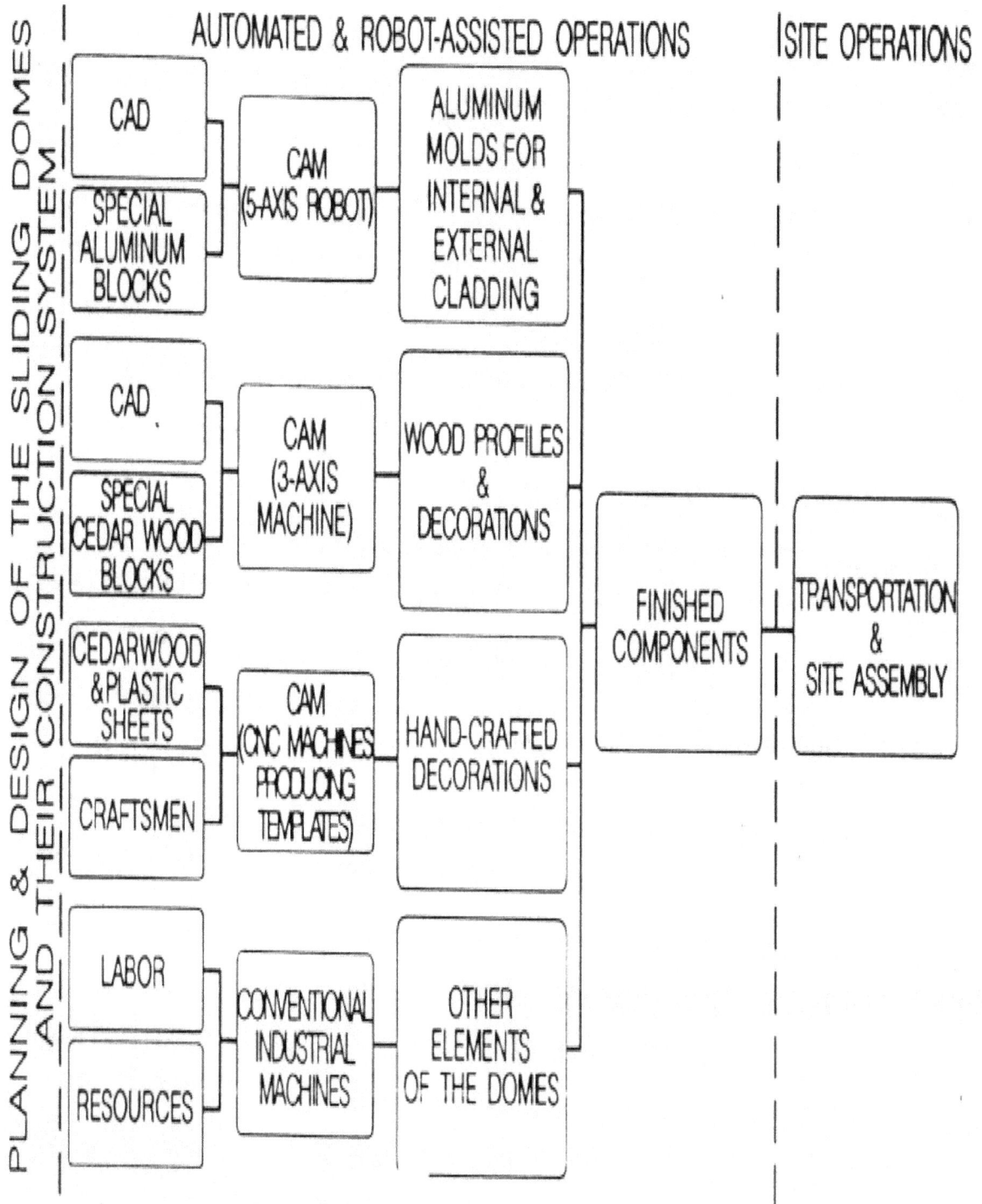

Figure 1: Robotics and Automation in Design and Construction of the Sliding Domes

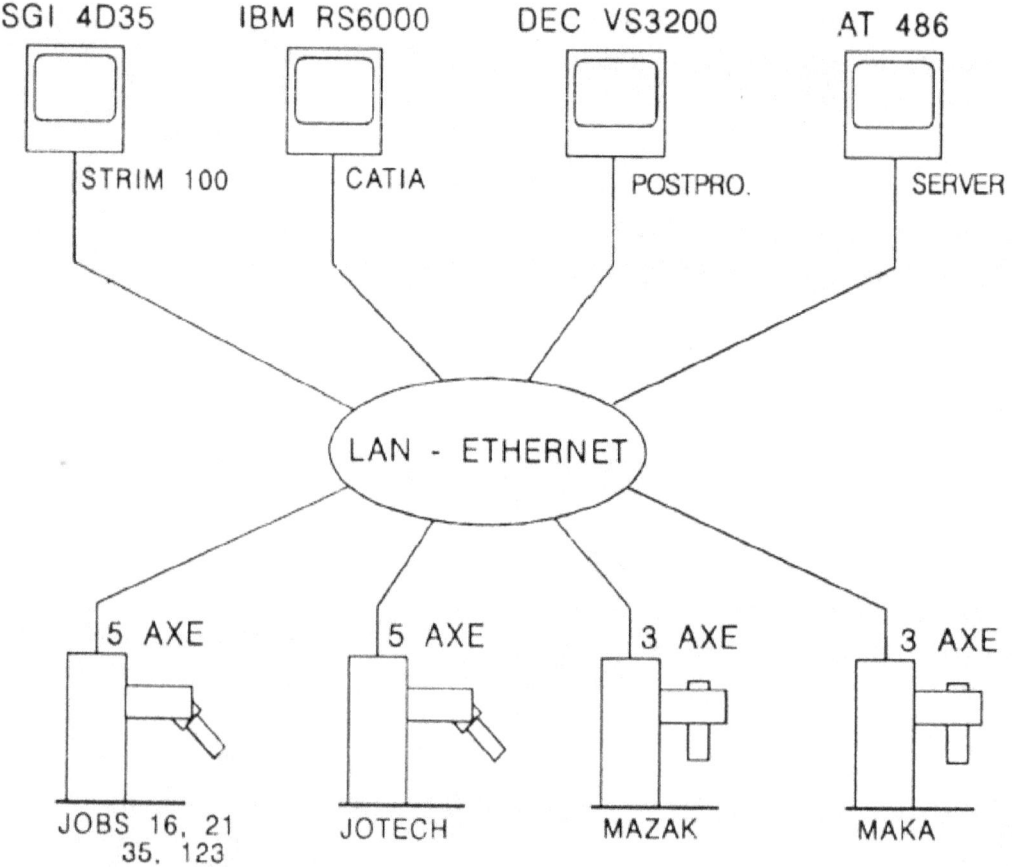

Figure 2: Interaction of CAD and CAM systems in the Sliding Domes Project

5.0 DIAGNOSIS

_ Equipment Diagnosis & Repair. Construction equipment is involving with repair and maintenance. These smart systems can maintain track of the equipment's internal conditions recognizing also systems as expert mechanics repair and preventive maintenance procedures.

_ Structure Diagnosis: these systems use as safety and serviceability of various classes of structure (bridges, buildings, dams, etc) would take data from tests and descriptions of a structure to assess.

6.0 QUALITY ASSURANCE AND PHYSICAL TESTING

The Hamersley Iron Organization in Western Australia has created a multi-functional, fully robotized, on-site material testing laboratory for the industrial quality control of bulk materials. The systems allow automatic breaking and sampling of bulk materials and the physical and chemical testing of samples. The remote on-site is controlled from a laboratory some 5 kilometers away. The system provides work 24 hours per day a week and data for quality control unit. This system robot is more economical and in comparison with manual testing.

7.0 MANIPULATOR SYSTEMS DEVELOPMENT

One application for this type of devices is in the destroy of concrete and industrial process structures but the system can be used in other applications such as sizing blocks , carrying of heavy parts or active loads at large reaches or on joining with other field manipulator system.

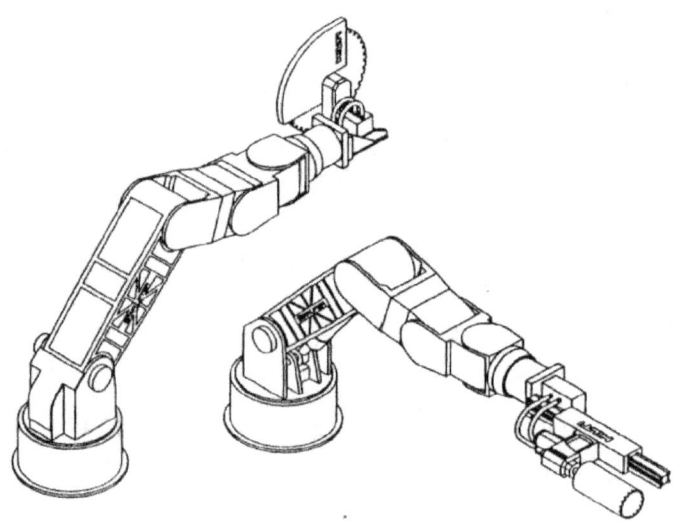

Figure 3: A robotic-action-head type electro-hydraulic manipulator for the deployment of heavy active-tool system

8.0 ASSEMBLING WALL SEGMENTS IN TUNNELS

Ishikawajima-Harima heavy industry and Tokyo Electric Power Company have jointly developed a robot to assemble wall segments in tunnels sewerage systems and power cables. In manual process this act to take three to four workers about 50 minutes to assemble a single ring of wall segments for tunnels four to five meters in decimeter. With the robot system the same operation can be performed in 30 minutes with only one labor operator. The operating sequence of the robot system is as follows. A special conveyor system feeds wall segments to an automatic assembly machine, the machine then moves each segment to the proper place and joints it with the bolt hokes by using a hole detecting sensor, the bolt tightener inserts and tightens the nuts and bolts.

9.0 STONE - STACKIING IN RESIDENTIAL CONSTRUCTION

Tokyo Construction Company has invented a stone – stacking robot designed for construction sites. The robot can carry heavy concrete blocks to an indicated place and install each block along the foundation. In manual process a stone or concrete block of about 60 kilograms is loaded onto the bucket and are carried to an unloaded at the construction site one by one by the workers. The robot hand can handle stone or concrete blocks of online shopping. These robots currently are used at housing construction site in Tokyo.

10.0 WELDING OF LARGE INDUSRIAL ELEMENTS

Mitsubishi Heavy Industries has produced a heavy- duty welding robot for welding large structural such as the panel blocks for cranes and bridges. This system can work five times the work capacity of human welders. The robots welding range is 58 meters wide and 32 meters long.

11.0 CONCLUSION

Today the potential for the evolvement and use of robotics in civil engineering is encouraging. At the moment, robotics and automation are the most important field in construction and civil engineering research in the United States and Australia. It does seem that, large companies and industrial groups should become active in increasing robotics application in civil engineering. Undoubtedly, one of the best ways for developing robotics in civil engineering is that, people change their mind about the house and life's place. The government should encourage people for living in the condominiums, skyscrapers, and producing building prefabricate.

REFERENCES

1. Daniel R. Rehak, ''Expert System in civil engineering, construction and construction robotics'' Carnegie Mellon University, showcase.

2. J.O Brian '' Construction Automation and robotics in Australia'' a state of a art view, Department of civil engineering, University of N.S.W, Sydney, Australia.

3. Jonathan O'Brian, ''Construction Automation and Robotics in Australia'' a state of the Art Report, school of civil engineering. University of New South Wales Sydney 2052, Australia.

4. Khalid A.H Taher, Bakr M.Binladin, Fuad A.Rihani, and Mahmud Rusch, ''Robotics and Automation in the construction of the Sliding Domes of King Fahd's Extension of the Prophet's Holy Mosque in Madinah, Kingdom of Saudi Arabia'' civil engineering Department King Saud University.

5. Miroslaw J. Skibniewski, ''current status of construction Automation and robotics in the United States of America'' The 9th International symposium on Automation and robotics in construction June 3-5, 1992 Tokyo, Japan.

6. Shankar Sundoreswaran and David Arditi The 5th International Symposium on robotics in construction June 6-8 Tokyo, Japan.

*****

Automated Robotic Maintenance for Industry

Nima baharaei

UPM University, Serdang, Malaysia

Academic Editor: *Amin Teyfouri*

ABSTRACT: The business model of high-value principal assets is shifting from buying a physical produce to obtaining a result or a function maintained by the product combined with a number of related services. One such service, maintenance, is perhaps the most effective way to keep the purpose available during the product development. Automation has played a dynamic role in industry during history, especially within the production line. With the movement towards providing product service systems the need for services such as maintenance are increasingly important for a manufactured product, and the pull towards automation may drive down costs and improve performance time. Though currently robotic applications to maintenance outside monitoring and inspection responsibilities are not usual, this research purposes at exploring the viability of future maintenance robots that can perform a variety of maintenance tasks. This work looks first at search, labeling and classification of a number of maintenance missions using standard industrial engineering techniques. This includes decomposing the maintenance work into a number of 'unit tasks' essential to be performed in instruction to complete the specified maintenance.

Keywords: Robotic Maintenance; Class 222 diesel engine train

INTRODUCTION

The business model for the provision of a wide variety of high-value capital assets such as aero engines, trains and medical scanners is experiencing a fundamental shift [1]. Previously the end user would purchase the asset from the manufacturer and become the asset's owner. Responsibility for the operability and maintenance of the asset would pass to the end user as soon as the manufacturer's warranty period had expired [2]. With a move towards a more service-based model, assets are now being leased out to the end user, with a contract with a maintenance provider to ensure availability of the asset [3]. This is seen on UK railways where the time-limited franchise for operating a particular service is given to a train operating company who then lease the rolling stock from its owner and contract with a maintenance company under a service level agreement to ensure availability of the asset.

This paper presents a novel methodology for the task classification of maintenance activities, with the final purpose in leading towards automation through robotic platforms.

MAINTENANCE AND ROBOTICS

Recent decades have seen a growing use of robots inside manufacturing processes. The rate, availability and improved accuracy of robots have had significant impact in reducing both manufacturing costs and through-life costs due to enhanced quality.

It therefore seems logical to discover the possibilities of robots being used within the maintenance function to provide the same benefits as are found by their use in the manufacturing process. However, so far applications of robots to maintenance have been limited to mostly monitoring and inspection tasks

[4]. Most of them are non-autonomous robots (e.g., remote-controlled) equipped with cameras, sensors, and master-slave controlled manipulators. Specially, a few robots are capable of performing their inspection tasks autonomously [5].

Maintenance is the process of keeping something in good order. The last purpose of this research is to examine the feasibility of introducing robots to maintenance beyond simple checking and inspection tasks. However, maintenance is identified to have a number of characteristics that are not found in the process of manufacturing of new products. Among others, maintenance is often irregular, non- uniform, non-deterministic and non-standardized.

INDUSTRIAL ENGINEERING TECHNIQUES AND CLASSIFICATION

IE techniques are into two fundamental categories; they are either used for analysis of existing processes or development of those processes. Reported uses of IE within maintenance activities tend to describe how techniques are used for the latter, either by increasing their efficiency or improving their planning. So the Single Minute Exchange of Dies (SMED) techniques are used to reduce setup times [6] and scheduling rules have been used to improve repair system performance within the system's existing constraints (e.g. labor hours, repair facility availability) [7].

As the focus of this research is the analysis of program, IE techniques that are designed for use in describing motion would appear most relevant. Consequently the time and motion study techniques developed by Frank and Lillian Gilbreth [8].

The Gilbreths identified 18 essential motions (to which they gave the name Therbligs) connected to motion by humans undertaking manufacturing processes. Therbligs are used to define time and motion with the system being protracted to record the program of each hand in a Simultaneous Motion (SIMO) chart [9].

On a higher level, maintenance tasks can also be recorded and defined using simple process flow charts to capture the main tasks and decisions within it.

TASK CLASSIFICATION METHODOLOGY

In order to fully classify a maintenance task so as to allow its eventual automation through robotics, a process is required to capture the necessary information. Discussed previously, within the literature there are a number of industrial engineering techniques which can be used to capture this information for a given task or function. Maintenance often involves complex motion and function and therefore the analysis of this movement is the focus of the developed methodology. A two-stage process is therefore outlined, with video recording of the maintenance event; this is recorded into a process flow, and then followed with more detailed analysis of the video to capture motion information through Therblig and SIMO classification. Figure 1 captures this process for maintenance task classification.

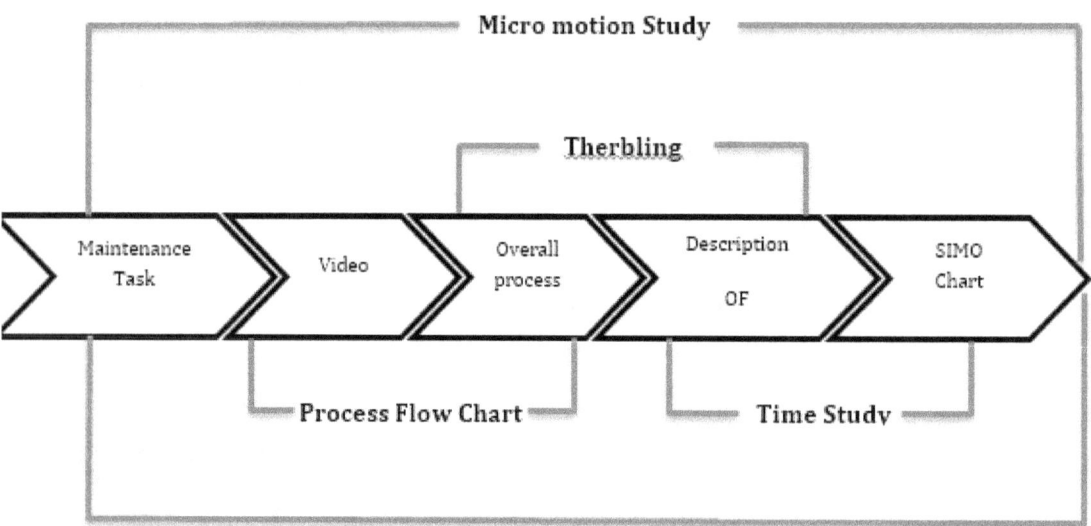

Figure 1: Task Classification Methodology

CASE STUDIES

In order to evaluate the proposed task classification methodology a number of maintenance case studies need to be performed. Outlined below are two separate case studies, the first focusing on a simple laboratory mockup using a radio controlled (RC) car, and then followed with an example of maintenance in the field on a class 222 diesel engine train undercarriage.

Radio controlled car

A petrol driven radio controlled car with over 10 years of heavy use provides a suitable platform for evaluating the proposed methodology. Here one task is chosen, the maintenance of a rear shock absorber, consisting of removal, inspection, cleaning and refitting. A breakdown of the generic steps required to perform this maintenance is shown in Table 1.

Table 1: Rear Shock Absorber Breakdown

Step	Description
1	Remove the 4 'R' pins, which hold the car body in place
2	Remove the car body to gain access to shock absorber
3	Remove shock absorber upper mount screw
4	Remove shock absorber lower mount screw
5	Remove shock absorber and inspect
6	Return shock absorber to car
7	Affix shock absorber upper mounting screw
8	Affix shock absorber lower mounting screw
9	Replace car body
10	Replace the 4 'R' pins, which hold the car body in place

Class 222 diesel engine train undercarriage

The class 222 is a diesel-electric multiple unit high speed train. Underneath each carriage on the train is a diesel engine, which drives an electrical generator, which then drives an electric traction motor. As well as driving the train, the traction motor provides rheostatic braking to reduce brake pad wears and in order to shield it from the elements a series of side skirts is employed. In order to perform maintenance it is necessary to lower and raise the side skirts to gain access to the traction unit and any ancillary equipment. A breakdown of the generic steps required to perform this maintenance is shown in Table 2. The electric traction motor in the undercarriage is also connected to a pair of bogey wheels through a final drive gearbox. One of the regular maintenance tasks on the gearbox involves checking and analyzing of the oil filler. A breakdown of the generic steps required to perform this maintenance is also shown in Table 3.

Table 2: Lowering and Raising Side Skirt Breakdown

Step	Description
1	Unscrew left and right hand securing bolt
2	Open latch and lower skirt to horizontal position
3	Remove spring clips on retaining lanyard from both sides
4	Lower skirt to vertical position
5	Raise skirt to horizontal position and re-attach spring clips
6	Raise skirt and close hatch
7	Screw in left and right hand securing bolts

Table 3: Checking Oil Filler Level Breakdown

Step	Description
1	Assemble sampling device
2	Remove filler cap
3	Visually inspect oil on spigot for contamination
4	Wipe spigot
5	Replace filler cap
6	Remove filler cap and check oil level point on spigot
7	Take oil sample
8	Replace filler cap

The next section looks at the experimental results achieved through the application of the task classification methodology outlined previously to each of the case studies.

EXPERIMENTAL RESULTS

The task classification methodology leads to the creation of a number of industrial engineering process and motion information of the maintenance task involved. For each case study maintenance example this includes video information, process flow charts, Therbligs and SIMO chart motion information. For both the RC car case study and class 222 diesel engine train undercarriage case study a number of results are shown.

Radio controlled car results

The disassembly of the RC car and subsequent maintenance is presented in Figure 5 as a process flow chart capturing the main steps of this procedure. Even with such a simple example the very first step highlights some of the problems faced, with removing the 'R' pins prior to the car body.

Here two hands are needed to undertake the task, one to hold the car stationary and the other to withdraw the pin. Consequently this would require either two general-purpose robotic arms working in collaboration or one robot with a special end effector, which can brace itself on the car before withdrawing the pin. In either case care has to be given in order not to damage the body structure when performing this task and detailed information is needed in order to perform it correctly.

The removal and replacement of the crosshead screws is another task in the process, which bears difficult challenges to robotic automation. In this instance both ends of the shock absorber are attached to the chassis using cross-head screws, often used due to their tolerance in the initial positioning of the screw driver when it mates with the screw head. The final application of axial force aligns the screwdriver with the screw head and provides a tight fit, which allows torque to be applied. This only works properly if the correct screwdriver is used, the screw-head has not been damaged previously and a suitable axial force is maintained during the screwing action (which in itself is a challenge as the screw head moves axially as it is rotated). If any of these conditions are not met then the screwdriver will 'jump out' and will quite likely damage the screw-head. This problem does not usually occur in the initial assembly of the item during the manufacturing process but there are a number of potential hazards in maintenance processes. Consequently it required some degree of dexterity together with tactile and visual feedback to carefully remove the screws without their being damaged in the process, for example thread stripping.

Table 4: Therbligs Step Two – Lift Off Body

No of Motion	Description	Therbligs
1	Search for the car body	Search
2	Find the car body	Find
3	Reach for the car body	Transport Empty
4	Grab the body with both hands (One on each side)	Grasp
5	Lift the body vertically until the aerial has passed entirely through the body	Transport loaded
6	Release the right hand	Release loaded
7	Search for the storage place for the car body	Search
8	Find the storage place for the car body	Find
9	Move the car body to the storage place	Transport Loaded
10	Release the car body alongside the RC car	Release Load

Figure 5: Process flow chart for shock absorber

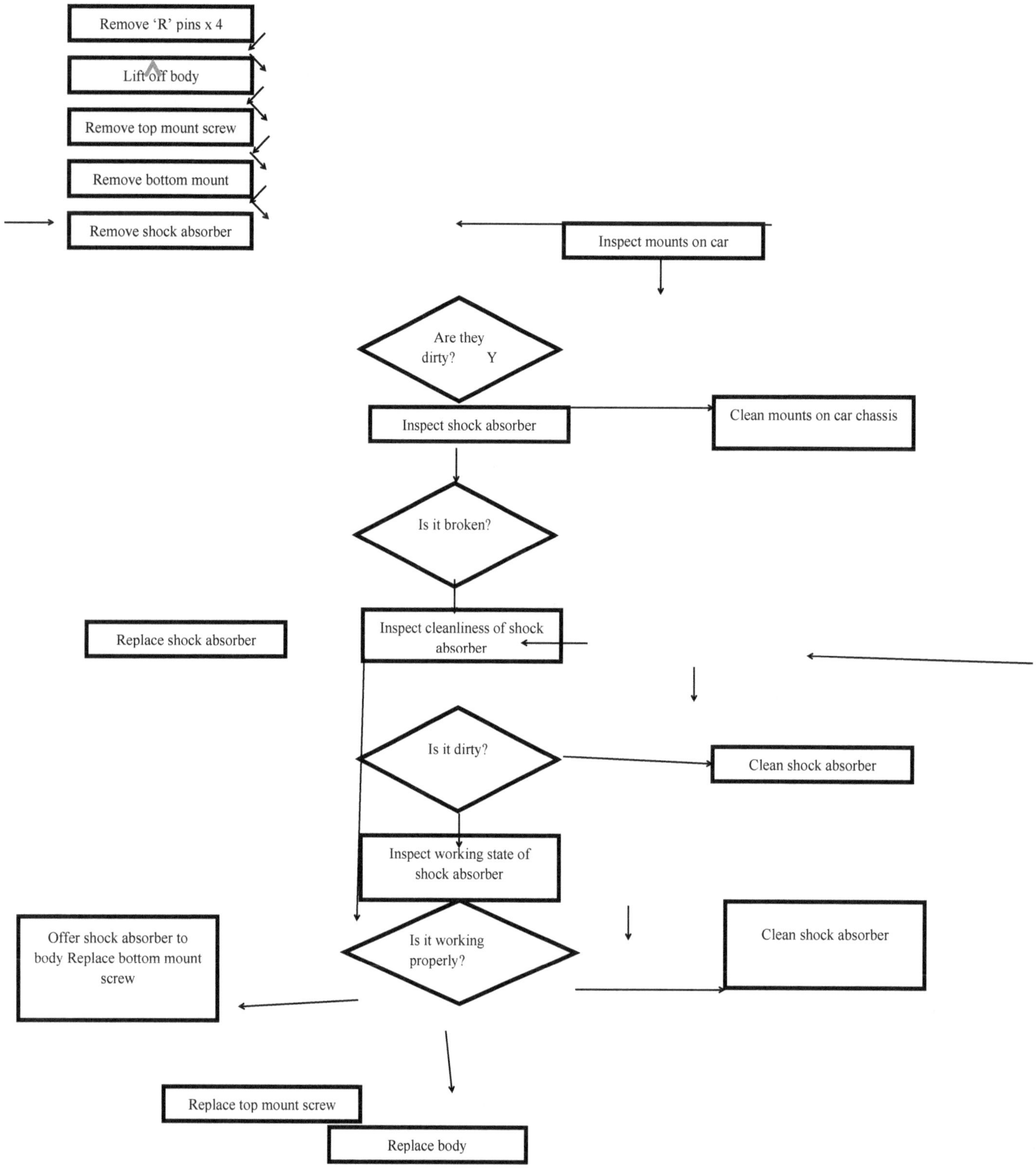

- Remove 'R' pins x 4
- Lift off body
- Remove top mount screw
- Remove bottom mount
- Remove shock absorber
- Inspect mounts on car
- Are they dirty? Y
- Inspect shock absorber → Clean mounts on car chassis
- Is it broken?
- Replace shock absorber
- Inspect cleanliness of shock absorber
- Is it dirty? → Clean shock absorber
- Inspect working state of shock absorber
- Is it working properly? → Clean shock absorber
- Offer shock absorber to body Replace bottom mount screw
- Replace top mount screw
- Replace body
- Replace 'R' pins x 4

Class 222 Diesel Engine   **Train Undercarriage Results**

The aluminum oil filler cap incorporates a magnetized steel rod with knurled markings that denote the maximum and minimum oil levels. It is retained in the cast iron gearbox casing by a bayonet fitting. Once the cap has been removed, the magnetic end of the rod is inspected for metal deposits before it is wiped and placed back into the gearbox casing and allowed to rest unfastened before being removed again and the oil level noted. It is then removed a second time to allow an oil specimen for debris analysis to be taken before being replaced once more and fastened into place.

A number of issues were identified during the process of oil filler maintenance. An oil-tight seal is produced because the elongated inside of the aluminum cap has a close fit to the accurately bored hole in the casing. Consequently the cap can be quite challenging to remove – usually requiring the shaft of a long screwdriver to lever it out. Bronze particles produced by wear of the brushes in the gearbox will not be attracted to the magnetized steel rod; however they will cause the oil to have an orange tint which can be identified by an experienced maintenance fitter. Similarly water contamination can be identified by sight without having to wait for the results of the laboratory analysis of the oil sample.

A number of lessons were also learnt from the skirt maintenance. Lowering the skirts is a two-stage process. Firstly the securing bolts at each end are unscrewed before the central latch is released which causes the skirt to drop. Its descent is arrested at the 90o position by wire lanyards at each end attached to the skirt by spring clips. Once these are disconnected the door can be fully lowered. Most maintenance is carried out with the skirt remaining attached to the unit but should it require completely removing, the hinge design is such that this can be achieved by simply lifting the skirt some 30mm.

Releasing and refastening the spring clips can be challenging and the lanyard is easily trapped when the skirt is raised. While using power tools to screw and unscrew the securing bolts may appear to save time, it can result in difficulty in unscrewing when the unit next comes in for maintenance.

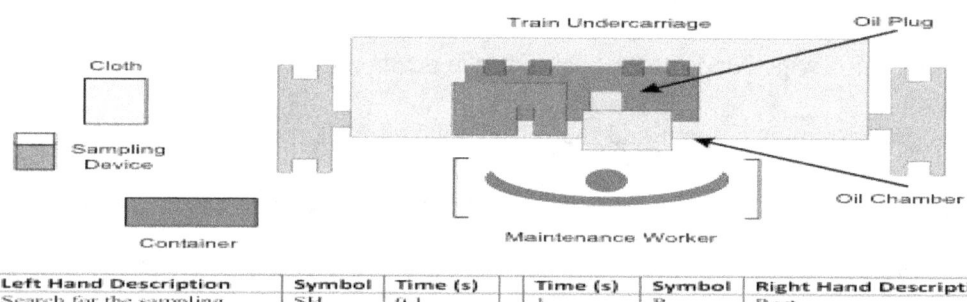

Figure 7: SIMO chart step seven – put the pipe in the oil

Table 5: Therbligs Step Seven – Put the Pipe in the Oil

No of motion	Description	Therbligs
1	Search for the sampling device	Search
2	Find the sampling device	Find
3	Grasp the sampling device	Grasp
4	Search for the pipe of the sampling device	Search
5	Find the pipe of the sampling device	Find
6	Grasp the pipe of the sampling device	Grasp
7	Move the sampling device to the oil chamber	Transport loaded
8	Move the pipe to the oil chambers hole	Transport loaded

9	Position the bottom of the pipe in front of the oil chambers hole	Transport loaded Position
10	Put the bottom of the pipe in the hole Release the pipe	Transport loaded
11	Release the pipe	Release

The latch on the heavy skirt is the same design as that on the lighter skirt. Consequently over time the latch on the heavier door is inclined to wear and ultimately fail. It is therefore imperative to support the weight of the skirt when unscrewing the second securing bolt in case the latch is incapable of keeping the skirt raised. An air filter assembly forms part of the skirt and should remain in place at all times. However with wear and tear, when the skirt is lowered to its final position it is not uncommon for this slide out and drop onto the floor. Should this happen, the assembly will need replacing and holding in place during the subsequent raising of the skirt. The weight of this skirt is such that when it is being raised the hinges are inclined to separate unless a second person presses them back into place.

DISCUSSIONS AND FUTURE WORK

The use of the chosen IE techniques can help to describe the process and motions of maintenance by decomposing the tasks into a number of unit movements. However there is often not enough detail to ascertain whether a robot can accomplish this task. Consider the task 'Lift off body' shown in Table 4. The Therbligs used for its description are Search, Find, Transport Empty, Grasp, Release Load, Transport Loaded and Release Load. These Therbligs describe the process quite correctly, but some details are missing. For example step 5, "Lift the body vertically until the aerial has passed entirely through the body" is quite easy to perform for a human being. However, several issues may occur if a robot has to carry out this operation. In fact, this activity is more complicated than just lifting an object. How can a robot know when the aerial has passed entirely through the body? What if the aerial gets stuck or blocked during the car body removal? The Therblig used for this motion – Transport Loaded – does not depict that complexity.

Another example of this decision-making issue is given from cleaning activities. How can it be decided or described when a surface is cleaned enough? At the moment, no IE technique can model and describe decision-making as an action.

It also appears that some Therbligs are too ambiguous. This is the case of "Use" for example. This is a general term to describe the motion to be done. In fact, if the operator is using a screwdriver, the way of performing should be more precise: should the rotation be clockwise or anticlockwise? What design of screw head is being used? Cross head, torx, slot, etc.? Or, considering another example, what should be understood when speaking about "using" a tissue to clean a surface?

So while Therbligs are helpful in visualizing the process, their focus on providing a structured methodology for calculating the time an activity should take to perform means they are not specific enough to describe exactly what is happening from a physical perspective. Therbligs such as "Transport Loaded", "Grasp" or "Position" do not give enough information about the required physical attributes to lift or move an object. They do not consider the questions "what is the required force to lift an object?" , "what pressure must be applied to grab an object?", "what is the speed, velocity or acceleration required to move an object?".

Similarly there are no Therbligs to take account of decision-making activities. During the maintenance on the train a coolant leak was observed. Therbligs would be needed to deal with the questions "What is to be done when a leak is found?" "Is the leak significant enough to require attention?", "where is the source of the leak?".

The IE techniques outlined in this research can describe and analyze processes but something is missing, especially if these techniques are used to convert activities performed by humans into activities to be undertaken by robots.

For a better description and analysis of a complex activity, more Therbligs are needed. For example the Therblig "Use" is too generic and some additional Therbligs should be created depending on the tool being deployed. For example if a screwdriver is used, Therbligs like "turn clockwise" or "turn anticlockwise" could be more appropriate.

In addition, more information to support the Therbligs are required, such as physical or geometrical parameters. For example:

- If a hammer is used the Therblig "Use" should be integrated with information about the angle to hit the part.
- If a fragile object is grasped, information about the pressure and the force to applied is needed.
- When a part is being lifted, if it is heavy, more force will be necessary
- The distance the object is being moved – is it within the physical constraints of the robot being considered for the activity?

Furthermore, Therbligs for decision making are needed to better answer questions like "what to do in this particular situation?", "is the task done properly?", "if something happens, what should be done?".

Referring to the train maintenance tasks performed, it is important to evaluate different situations using all the different senses. For example, it is crucial to recognize an oil leak from the smell or the sight of drops or spots. Other senses involved are hearing, to notice suspect noises, or touch to detect heat or smoothness of surfaces.

It was asserted in section 2 that key characteristics of maintenance are that it is irregular, non-deterministic and non- standardized. This was demonstrated during the experimentation on the train. The results of the oil analysis from the final drive could require that unplanned maintenance be undertaken, removing the filler plug from the final drive often requires extra assistance from a large screwdriver and the work required to lower and raise a skirt varies depending on whether it is heavy or light.

CONCLUSIONS

An investigation into the role autonomous robotics can play in industrial maintenance was undertaken in this research. Focusing firstly on developing a maintenance task classification methodology using known industrial engineering techniques and then validating it against two separate case studies, a radio controlled car shock absorber and a class 222 diesel engine train undercarriage. The results point to a successful implementation but also a need to further improve the task classification process. Current IE techniques were unable to provide the suitable detail needed for application of autonomous robotic platforms in this case.

Future work will look at improving this methodology further, and looking at developing a physical autonomous maintenance demonstration based upon this future maintenance task classification.

REFERENCES

[1] Baines TS, Lightfoot H, Steve E, Neely A, Greenough R, Peppard J, Roy R, Shehab E,

Braganza A, Tiwari A, Alcock J, Angus J, Bastl M, Cousens A, Irving P, Johnson M, Kingston J, Lockett H, Martinez V, Michele P, Tranfield D, Walton I, and Wilson H. State-of-the-art in product-service- systems. Journal Engineering Manufacture 2007;221:1543-1552.

[2] Takata S, Kimura F, van Houten FJAM, Westkamper E, Shpitalni M, Ceglarek D, and Lee J. Maintenance: changing role in life cycle management. Annals of CIRP 2004;53:643-655.

[3] Zhang Z, and Chu X. A new approach for conceptual design of product and maintenance. International Journal of Computer Integrated Manufacturing 2010;23:603-618.

[4] Parker LE, and Draper JV. Robotics Applications in Maintenance and

Repair. 1998. Handbook of Industrial Robotics, 2nd Ed.

Ultrasonics 2011;51:258-269

[5] Cakmakci M. Process improvement: Performance analysis of the setup time reduction-SMED in the automobile industry. International Journal of Advanced Manufacturing Technology 1995;41:168-179.

[6] Dutta U, and Maze TH. Application of Industrial Engineering Techniques in Transit Maintenance. ITE Journal 1987;57:45-49.

[7] Gilbreth FL. Motion Study. Princeton, N.J: Van Nostrand 1911.

[8] Price B. Frank and Lillian Gilbreth and the Manufacture and Marketing of

Motion Study. 1908-1924. The Business History Conference Edn.

[9] Dobie G, Spencer A, Burnham K, Pierce GS, Worden K, Galbraith W, and Hayward G.. Simulation of ultrasnoic lamb wave generation, propogation and detection for a reconfigurable air coupled scanner

*****

Managing the Chain of Suppliers

Shahryar Sorooshian[1], Ali Asghar Jomah Adham[2]

[1] *Department of industrial technology management,, Faculty of Technology, University Malaysia Pahang, Malaysia*

[2] *Faculty of Technology, University Malaysia Pahang, Malaysia*

Academic Editor: *Siti Aissah Mad Ali*

ABSTRACT: Demand for better product diversity and increased worldwide competition effect in the emergence of dynamic supply chain, which need to be work integrated with all parts in the chain without giving up their flexibility. Lack of Flexibility, deficiencies of traditional application systems such as MRP I, MRP II and ERP and need to have self-sufficiency cause to apply a dynamic network. In addition, EDI interface ignore the decision values and also EDI is very expensive and hard logistic handling required for supply chain management. This study is intended to survey traditional and modern adaptable software and applications, which have to be employed by enterprise for integrating across organizational network, by given the limitations and advantages of each application systems. Finally, the role of internet in supply chain is determined and modern way consists of e-commerce, e-supply chain and web-based supply chain are investigated.

Key Words: e-supply chain, e-commerce, web-based supply chain, Information Technology

1.0 INTRODUCTION

The aim of this study is figuring out a special trend in SCM performance from traditional methods to advanced and electronic based implementation. Also insufficiencies of traditional and modern methods that are using in SCM are analyzed. Finally, we discuss about SCM under electronic and web-based environment. According to previous studies, traditional supply chain performance ignores the whole system measurement of supply chain and it focused only on single company or department attitude. It means that each department is considered as an independent unit for setting performance standards and goals individually. On the other hands, nowadays, competitions among enterprises move between their supply chain and therefore, quick reaction will be needed. In addition the traditional SCM performances aren't able to reflect the creative capability of enterprise, dynamic condition, and continual improving ability and SCM performance because information of traditional evaluation is based on financial results. So, the evaluation system should be changed and transforming from the cost-orientation to the market-orientation is necessary for organizations.

Similarly, traditional SCM performance evaluating ignores the relations between enterprise and its external advantage relaters.

Indeed, it stress only on inner evaluation and pays its attention just on the inner procedures.

Therefore, this systems will lost their good agility and high efficiency [21].

2.0 METHOD

Applying information technology in supply chain management follows the special trend that is investigated in this study. To achieve the target we review previous researches that studied on this field. Indeed, the literature review is employed in this research. A literature review is an organized, explicit and renewable devise for evaluating, recognizing and realizing the main of existing for recorded analyze of all related papers in the special field by means of a structured documents [15]. Literature reviews usually intend at two purposes [18]:

1. Existing research is summarized by considering to subjects, issues and patterns identification.

2. contributing to theories by helping to classify the conceptual content of the field.[209.17]

The main aim of analyzing documents is opening up data or materials that don't need to be produced based on data collection by the researchers. In addition, providing complete reviews might be possible only with emerging or narrowly defined problem.

For assessing structural content criteria while quantitative and qualitative characteristics are mixed, literature reviews are be able to figure out as content analysis [201-20]. Mayring has proposed the process model including four steps[214-19]:

1. Material collection

2. Descriptive analysis

3. Category selection

4. Material evaluation

3.0 BACK GROUND OF SUPPLY CHAIN AND SUPPLY CHAIN MANAGEMENT

Supply chain is the sequence of firms and network of facilities, technologies, information, functions, people and resources, which includes suppliers, producers, warehouses, distributors, transporters, vendors and customers [13]. In manufacturing or business activities, the transmission between information flow, money flow, business flow and logistics flow is generally according to supply chain management (SCM) [22]. Also SCM includes designing and managing all functions within chain from finding resource and supplier to all aspects, which are related to logistics and delivery, finished goods to final customer. On the other hand, with the change of customer's behavior and demands, marketplaces require more and more customized products, but in this area, the traditional supply chain management has three essential disadvantages. First, low forecast accuracy because of uncertain demands leads to increase amount of inventory by producers and vendors. Second, traditional supply chain doesn't have enough flexibility and quick respond to demand changes, as an instance when sales is increased unexpectedly, traditional supply chain couldn't react easy and fast. Finally, the third weakness is about losing opportunities to reduce cost because TSC has same strategies to different items regardless to their volume.

Supply chain management consists of three major procedures [5] which consist:

1. Information Management

2. Logistic Management

3. Relations Management

It is essential to recognize that one of the most important requirements to reach successful SCM is the integration of information flow, which starts form Cooperation, next coordination and finally collaboration [10].

Information sharing has significant effects on all supply chain processes and SCM integrating but most of information is unnecessary, often late and frequently accurate [10]

3.1 Information sharing in supply chain

Information play important roles in supply chain efficiency. Correct and on time information have a significant effect on speed and excellent of decision-making and logistic section. In fact, information sharing influenced all supply chain aspects such as availability supply volume demand data better customer services with superior quality, improving collaboration between suppliers and optimizing the entire supply chain by offering better forecast do to reduce inventory. Actually, in today industry world, competition is between sets of enterprises and their supply chain [2,11].

By development of information technology and internet companies are able to share and transfer data in real time. In addition, it enables many of firms to expand flow of information for sharing skills, capabilities and knowledge regardless of location. Indeed the nature of information management and information transferring has modified. On the other hand lack of commercial links among the enterprises and no trust relationship between supply chain partners without sharing data network, led to disproportionate and chain interruption [12].

3.2 Information Technology in supply chain

In today's technology-demanding world, companies figure out the use of Information Technology (IT) is one of the most important issues for monitoring complex global chains. Supply chain management intends to control cohesive network of multi-companies due to provide impressive profit, so IT is a fundamental prerequisite.

Before 80s, partners of supply chain disregarded most of the time information process. Actually, it was applied as theory and was not very important and before60s, decade information processing was like sheet.

Prevent interference in the supply chain and reduction of differences in transaction between various sections or persons and its purposes in supply chain includes [9]:

- Visibility and availability of information
- Information promoting with all aspects of contact points
- Ability for decision-making based on information for entire supply chain

Organizations apply hardware such as computer and software including ERP sources, OPT optimization and SCM management to develop and maintain information technology usage in supply chain.

By regarding to other types of divisions, three main outcomes, which companies use informational technology are found according to figure 1.

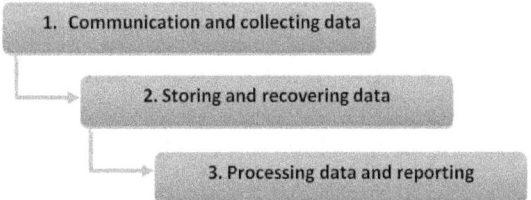

Another way that IT enables whole supply chain to integrate essential activities and automate process is an Electrical structure. Information Technology has been applied in supply chain management by two main categories consist of ERP software group and Internet supporting.

4.0 BACKGROUND OF ERP SOFTWARE GROUP

The most common technology, which can be used in all types of enterprises, is ERP and ERP software group.

4.1 MRP (Material Resource Planning)

Before the advent of computers and prior to MRP, many of processes have been done by paper. In 1960s, IBM presented some concepts of MRP and Joseph Orlicky developed Material Resource Planning by investigating on TOYOTA manufacturing program. The main three scope of MRP includes inventory control, basic programming and BOM programming. MRP is able to control levels of inventory as well as schedule manufacturing, purchase and transportation activities, in spite of considerable MRP benefits, there are many important gaps in this system. As instance, the major problem with MRP is the strong dependency to accurate information. It means that output data will be wrong if there are any mistakes in inventory or BOM data. Another problem with MRP system is necessity to very long cycle lifetime, expensive computers, ignoring relationship among organizations and supplier and above all discards competitive strategy in supply chain management and capacity in its calculations. This means it will give results that are impossible to implement due to labor, machine, or supplier capacity constrictions. Consequently, it was replaced by MRP II.

4.2 MRP II

In 1980s, although enterprises were able to designing, scheduling, inventory control and cash fallow with MRP II, it was highly dependent with integrating all features of operational levels. MRP II cannot optimize system applications. In addition, it loses its optimal applications when productions have different sequence of processes. MRPII was aimed to integrate a lot of information by centralized database. However, in 1980s the hardware, software, and relational database technology was not advanced enough to provide the speed and capacity to run these systems in real-time, and implementing these systems was too costly for most businesses. Then ERP has been the next step.

4.3 ERP (Enterprise Resource Planning)

In 1990s, ERP with aim of integrating all section, resources and activities was expanded rapidly. In that time, ERP systems focused on functions that were not related to final customer but the end of 1990s it begun to expand in supply chain activities due to integrate relationship among enterprises, suppliers and customers. Although ERP system increases customer satisfaction, it has never considered supporting entire supply chain. It just focuses around business satisfaction similarly ERP system mechanisms is not based on actual workshop that it influences on lead-time and operation activities considerably. In spite of substantial abilities of ERP system to access information from different parts of organization, there is no potential to explore or expand owing to shortage of cost availability common standard for supply chain. Finally, in most of ERP systems have been already used only local site are able to access data or supply chain information and network functions for purchasing. It couldn't perform as global systems for external suppliers. In 2000s, ERP II was invented. It is about web-based software. It means that both customers and suppliers will be able to access system and information in real time.

The next way that information technology is involved in supply chain includes internet.

5.0 INTERNET IN SUPPLY CHAIN MANAGEMENT

Internet offers tools for SCM that can be used to reduce costs and cycle time, increase customer satisfaction, develop dynamic pricing and production coordination. Another words internet equips supply chain management process to be performed in expediting maximum supply chain implementation, immediately manner and harmonizing. Electronic supply chain management is not a new theory because many companies have already used electronic commerce applications such as classical EDI [4]. Using internet in supply chain management will be described in three categories consist of E-commerce, E-supply chain and web-based supply chain.

5.1 Background of e-commerce applications in supply chain management

E-Commerce means that means that companies apply internet due to deal all business activities and transaction through the world [14]. In fact, essence of supply chain management is changed by E-Commerce and Internet in order to improve cost efficiency, distribution systems, speed operations, costumer orientation and added-value whole supply chain partners[25].

E-Commerce technologies are very appropriate to provide tow important technological requirement for successful SCM on sharing information including first, Electronic links with high security and stability between companies and second, high broad-brand and integrated environment, also these requirements have an internal roles to create and equip new shape of SCM [13]

Another word E-Commerce enables supply chain to remove unnecessary process and joint and integrate higher down-stream investment. Figure 2 illustrate that transforming organization from traditional 4P to modern 4C by E-Commerce.

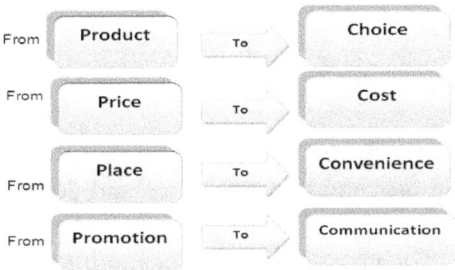

Figure 3: From Traditional 4P to Modern 4C

One of the common types of e-commerce and B2B networks is EDI that indicates the computer-to-computer exchange of regular commerce data between trade partners in standard data formats. The EDI standard says which pieces of information are compulsory for a particular document, which pieces are optional and give the rules for the structure of the document. The standards are like building codes.

5.1.2 Electronic Data Interchange (EDI)

One of the traditional features of e-commerce is EDI. EDI is a shape of inter organizational electronic business where one trading partner side including a buyer or seller establishes individual links with one or more trading partners through a computer-to-computer electronic communication method [23]. In fact, it is a common technological approach to exchange information and general business data among enterprises, which was used in 1980s [24] on a large scale of railway operations management for the first time [6]. Controlling and improving EDI has become a one of key management issue among IS executives[7] innovations on the Internet are creating new EDI communication qualifications. By considering to the large-scale connectivity and few geographical constraints, the internet emerges to be the most qualified stage for EDI [3]. Internet-based technologies, such as Web services software and standard TCP/IP communication protocols, stimulated growth of the intranet as a communication platform for standardized electronic information between businesses. This shows the requirement to understand the features that facilitate successful implementation of EDI within organizations.

Although EDI enables supply chain to improve customer service and delivery, reduce paper works, increase productivities and cost efficiency, it involves some important deficiencies. As an instance, applying EDI with internal computer systems of organizations and their trading partners is significantly difficult and very expensive. Also EDI is be able to focus on data exchanging only therefore supply chain management miss the decision value. Moreover, security issues such as inability to prevent of message modifying classification and contents, disclosure of message and sender, are another lack of EDI. In addition EDI development is to challenging and it con not be used in wide sprat and it is the most important deficiency of EDI.

Therefore, enterprises have became enthusiastic to use of internet with wide sprat in supply chain and avoid disadvantages that mentioned above. To apply e-commerce to the management of supply chain creates a new hot word that is e-supply chain management (e-SCM). E-SCM is to deal with organizations' information, resources and productss efficiently and effectively, and run information flow, capital flow and logistics smoothly during their suppliers, their vendors and their customers [4].

5.2 E-supply Chain Management

E-SCM is an internet supported supply chain to manage information and making logistic flow more effectively. Amount of inventory, labor and organizational cost is reduced by E-SCM. E-SCM not only has a good consideration on information sharing but also it concentrates on well-organized relationship between companies, suppliers, purchasers and customers.

Sustainable Competitive Advantage (SCA) will be built much better by applying E-SCM and quality of delivery service, transportation and productivity will be improved as a same time.

Moreover, data transforming will be done much faster. In addition, knowledge management adopts both of technology and of human managing in E-SCM system. It means that E-SCM system involves technology attitudes (Web-based and using boar-brand) and performance (sharing activities). e-SCM systems can be arranged from simple works such as an electronic catalog is put on an e-marketplace to complex jobs as a platform to enable product design and development between an entire supply chain .

it is so clear that enterprises should carry out some e-SCM strategies in their management with attention to integration of the supply chain [4]. These strategies are divided to three parts:

1. Managing of Suppliers and customers through e-commerce
2. Managing Market requirement forecast through e-commerce
3. Outsourcing management through e-commerce

5.3 Web-based Supply Chain Management

The Internet has developed into a widespread and professional communication tool in supply chain management, therefore Web exchanges have also emerge as a feasible alternative that in supply chain. So many companies are Enthusiastic to employ web-services throughout their supply chain [8]. Because of advantages of the new medium, many firms have attempted to take it.

For sharing the information that is used in supplier evaluation, enterprises have adopted software and hardware and have linked up with their suppliers on web exchanges. Many web exchange companies take place during the dot-com era in hopes of becoming providers for both simple and complex-product industries [1,7]

6.0 DISCUSSION AND CONCLUSION

From the above discussions, we found that throughout e-SCM, organizations are be able to manage suppliers, vendors and customers, organize the outsourcing, estimate the potential market demand more accurately and control the stores in systems that are more proficient. In this study, advantages of implementing e-SCM in business and many of its profits are explained and traditional applications are compared with modern applications. By applying internet and web-based systems, both small-sized and medium-sized enterprises can come into the global supply chain with low cost. What's more, B2B e-SCM is the direction of e-SCM. In the future, the collaboration between organizations will be the dominating and leading satisfied of the e-SCM.

REFERENCE

1. A.J. Senn. (1998), Expanding the reach of electronic commerce: the internet EDI alternative, Information Systems Management

2. Bazerman, M.H. "Conducting Influential Research: The Need for Prescriptive Implications," 2005.

3. Arcan Arsan and Aimee Shank. (2011), Associated Technologies: An Analysis of Supplier Evaluation. SCRC Article Library.

4. Chen Hanlin. (2007), Implementing E-Supply Chain Management in Enterprises — a Case Study. IEEE pp. 1-4244-1312.

5. James B. Ayers, (2000), "Supply chain Management".

6. Johnston, R. & Mak, H. (2000). An Emerging Vision of Internet-enabled Supply Chain Electronic Commerce. International Journal of Electronic Commerce, Vol.4, No. 4, pp. 43-59.

7. J.C. Brancheau, B.D. Janz, J.C. Wetherbe. (1996), Key issues in information systems management. SIM Delphiresults, MIS Quarterly, Vol. 20, No. 3, pp. 225–242.

8. Jizi Li, Ling Yuan and Jun Guo. (2009), Business Integrated Architecture for Dynamic Supply Chain Management with Web Service. International confrance on new trends in information and service science. Vol.6, No. 4, pp. 356-361.

9. Min Zhang; Hui Xiong; Jing Li. (2009), Optimization patterns of on-value electronic supply chain integration strategy

10. R. Michael Donovan. (2009), e-Supply Chain Management: Prerequisites to Success. PART I. Performance Improvement

11. Sepide Tavassoli, Maryam Sardashti and Naghme Khajeh Nasir Toussi. (2009), Supply Chain Management and Information Technology Support. IEEE, pp.289-293

12. Suhong Li and Binshan Lin. 2006. Accessing information sharing and information quality in supply chain management. Science direct. Vol. 16, No. 3, pp.1641–1656.

13. Xingxin Gao, Zhe Xiang, Hao Wang, Jun Shen, Jian Huang; Song Song. (2004), An approach to security and privacy of RFID system for supply chain. E-Commerce Technology for Dynamic E-Business. International Conference. pp. 164 – 168

14. Zhang Yuanyuan, Wang Ying, Qu Jiabin. (2010), Research on Supply Chain Management based on E-Commerce. Industrial and Information Systems (IIS), 2nd International Conference.

15. Fink A. Conducting research literature reviews: from paper to the internet. Thousand Oaks: Sage; 1998.

16. Harland CM, Lamming RC, Walker H, Philips WE, Caldwell ND, Johnson TE,

17. Supply management: is it a discipline? International Journal of Operations & Production Management 2006; 26(7):730–53.

18. Meredith J. Theory building through conceptual methods. International Journal of Operations & Production Management 1993;13(5):3–11.

19. Mayring P. (, Qualitative Inhaltanalyse – Grundlagen und Techniken. [Qualitative content analysis]. 8th ed. Weinheim, Germany: Beltz Verlag; 2003.

20. Dyllick T, Hockerts K. Beyond the business case for corporate sustainability. Business Strategy and the Environment 2002;11(2):130–41.

21. Xing Guo-Zheng, Jiang Yu-yan *, Sun Hao, Wang Zheng and Zhang Zhu-ting. (2011). The Supply Chain Management Performance Evaluation under Web-Based E-Commerce Environment. Energy Procedia. No (13). pp. 4020-4025

22. Lanqing Liu. (2011). Research on the Management System of enterprises using Modern Logistics Supply Chain Theory International Conference on Advances in Engineering. Energy Procedia. No. 24. pp. 721-725.

23. Narayanan, S., Marucheck, A.S., Handfield, R.B. (2009). Electronic data interchange: research review and future directions. Decision Sciences 40 (1). pp.121–163.

24. Emmelhainz, M.A., 1990. Electronic Data Interchange: A Total Management Guide. Van Nostrand Reinhold, New York.

25. Jouni Kauremaa, Juha-Miikka Nurmilaakso, Kari Tanskanen. (2010) E-business enabled operational linkages: The role of Rosetta Net in integrating the telecommunications supply chain. Int. J. Production Economics. No. 127. pp.343-357.

*****

Proposal of Study on Determinants of Consumer's Purchasing Intention towards Green Products

Lee Jo Ying[a], Liu Yao[a], Lee Chia Kuang[a], Chong Kwok Feng[b]

(Faculty of Technology[a], Faculty of Industrial Science & Technology[b], Universiti Malaysia Pahang, Tun Razak Highway, 26300 Kuantan, Pahang, Malaysia)

Academic Editor: *Shahryar Sorooshian*

ABSTRACT: Nowadays, consumer's purchasing behavior has exerted huge impacts to the environment. As the environment continues to deteriorate, Malaysia government is taking steps in promoting the green movement nationwide. To better understand the consumer's purchase of green products, it is imperative to know their purchasing intention. Hence, this study is aimed to investigate the determinants of purchasing intention towards green products among Malaysian consumers. Based on the Theory of Planned Behavior, the attitude, the subjective norm, and the perceived behavioral control (i.e. control on availability and perceived consumer effectiveness) are set as independent variables to be used to predict the dependent variable 'purchasing intention towards green products'. Survey questionnaire will be designed and employed to collect data from 200 Malaysian consumers. Quantitative analysis will be conducted using SPSS 20. The results of this study are expected to ascertain the relative strength of each determinant and to show positive relationships between each independent variable and the dependent variable. Hopefully, this study could be served as guidelines for green entrepreneurs who are planning to target the Malaysia market, as well as policy makers.

Keywords: Purchasing Intention ; Green Product ; Theory of Planned Behavior

1. **INTRODUCTION**

Green business is becoming trendy in Malaysia. Recently, the Malaysian Government is actively involved in green projects, namely in green technology, encouraging green consumerism and promoting green business among Malaysians. According to the World Environmental Performance Index (EPI), Malaysia was ranked at No. 27 out of 163 countries in 2008 but declined to No. 54 in 2010 [1]. And in 2011, Malaysia was ranked the 25th out of 132 countries [2], which is the best ranking for Malaysia in the EPI thus far and Malaysia is still striving to improve its ranking on the way to achieve a fully developed country in the year of 2020.

Green product can be defined as a product that is nontoxic, made from recycled materials, or minimally packaged [3]. Green products are based on few characteristics, including original grown, recyclable or reusable, containing recycled content, no pollution to the environment, approved by chemical test and on non-animals [4]. Thus, these environmentally sustainable products can help to reduce the environmental impact caused by the usage of harmful products. However, the practices of environmental responsibility are still low among Malaysian consumers even though Malaysia has performed enormous development followed by governmental efforts to attract foreign direct investment.

Consumers who are environmentally conscious do not always purchase the green products [5]. Furthermore, They are unlikely to engage in pro-environmental behavior if they think their action would not contribute to positive environmental result [6]. On the aspect of eco-friendliness of behaviors and habits, only 8% of Malaysian respondents responded that they had changed their behavior in a great deal to benefit the environment through a survey conducted online by the global market insight and information group TNS in 2008 [7].

Consequently, it is essential to assess and change the relevant human behavior in order to improve environmental sustainability [8] since their purchase behaviors largely affect the environment. And since the purchasing intention is a critical factor to predict consumer behavior [9], the factors that influence the purchasing intentions towards green products among Malaysia consumers would be worth to find out. Additionally, the environmental sustainability marketing studies in Asian countries are relatively less compared to Western countries [10]. So, it is vital to conduct a study in the context of Asian country such as Malaysia to understand their purchasing intention towards environmentally sustainable products. And this will be beneficial for the green policy makers and marketers to adjust or amend strategies, products as well as services.

Thus, the purpose of this research is to ascertain the determinants of purchasing intention towards green products among Malaysian consumers by incorporating the framework of the Theory of Planned Behavior. Specifically, this research aims to: i) to examine the relationship between attitude and the purchasing intention towards green products among Malaysian consumers; ii) to investigate the relationship between subjective norm and the purchasing intention towards green products among Malaysian consumers; and iii) to identify the relationship between perceived behavioral control and the purchasing intention towards green products among Malaysian consumers.

2. LITERATURE REVIEW AND RESEARCH FRAMEWORK

The Theory of Planned Behavior (TPB) proposed by Ajzen provides a complete framework to explore the influencing factors of behavioral decision [11;12]. According to it, the three independent determinants of intention consist of attitude, subjective norms and perceived behavioral control [13]. The attitude towards the behavior relates to the degree to which a person has favorable or unfavorable evaluation or appraisal of the behavior in question. Consumer's attitude is expressed as a composite of a consumer's belief, feelings and behavioral invention [14]. The subjective norm refers to the perceived social pressure to perform or not perform the behavior. The perceived behavior control represents the perceived ease or difficulty of performing the behavior and it is assumed to reflect past experience as well as anticipated obstacles. In this study, the control on the availability of environmentally sustainable products and the perceived consumer effectiveness are taken into consideration to explain the perceived behavioral control.

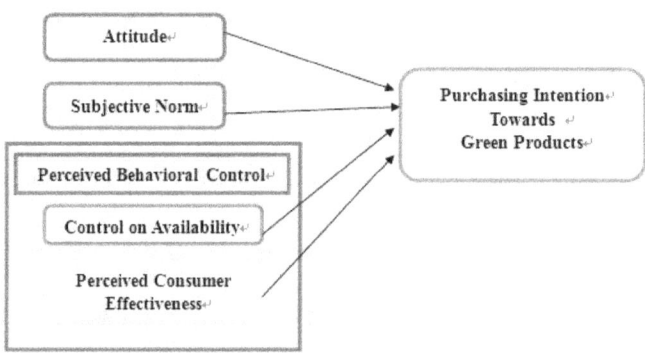

Figure 1: Research Conceptual Framework

The control on the availability of a product is the degree of difficulty or ease in locating and gaining a product for consumption. An individual's confidence is concluded in his ability to control displays a positive relationship with the purchasing intention or the purchasing behavior [15]. The perceived consumer effectiveness of a product is the persuasion that the individual has the ability to manipulate the result in a positive manner as a result of their action. The perceived consumer effectiveness is indicated to have a significant relationship with the perceived behaviour control [16]. Figure 1 shows the conceptual framework for this research, which consists of four variables. Purchasing intention is served as the dependent variable. The independent variables of the attitude, the subjective norm, the control on availability, and the perceived consumer effectiveness are supposed to be the determinants of purchasing intention.

3. RESEARCH HYPOTHESIS

According to the research framework and previous research, the research hypotheses of this study are formulated as follows.

H1: The Attitude has a positive relationship with the purchasing intention towards green products among Malaysian consumers.

H2: The Subjective norm has a positive relationship with the purchasing intention towards green products among Malaysian consumers.

H3: the Control on availability of the green products has a positive relationship with the purchasing intention towards green products among Malaysian consumers.

H4: the Perceived consumer effectiveness has a positive relationship with the purchasing intention towards green products among Malaysian consumers.

4. INSTRUMENTS AND DATA COLLECTION

To achieve the research objectives, questionnaire will be constructed using Microsoft Word and administered through an online survey via email and Facebook to a convenience sample consisting of 200 Malaysian consumers including university undergraduates and postgraduates as who are considered the future consumers having ability to transform the consumption pattern in the coming years. Using student sample is also regarded valid for exploratory studies [17].

The questionnaire under research will be designed using both open-ended and close-ended questions. There are mainly two sections: Section A and Section B. For Section A, nominal scale is chosen to collect the demographic information of the respondents. The respondents are required to tick the appropriate boxes and fill in their personal details in the provided blanks. While, Section B covers the independent variables and dependent variables. The established validated items that will be used for measuring attitude, subjective norm, control on availability, perceived consumer effectiveness, and purchasing intention are taken from the extant literature relevant to environmentally conscious consumer behavior. The 5-point Likert scale will be used for measuring the responses , where 5 denoted strongly agree, 4 agree, 3 neutral, 2 disagree and 1 strongly disagree. Table 1 demonstrates the questionnaire items that will be constructed based on the source given.

Table 1: Construct Questionnaire Item

Construct	Items	Sources
Attitude	4	[18]
Subjective Norm	4	[19]
Control on Availability	3	[20]
Perceived Consumer Effectiveness	4	[21]
Purchasing Intention for Green Products	4	[22]

5. DATA ANALYSIS AND INTERPRETATION

Statistical Package for the Social Sciences (SPSS) software will be used to analyze the collected data from all the filled questionnaires. Frequency, reliability and correlation will be conducted through SPSS sequentially. The findings will be presented accordingly in tables and graph forms to ease the process of interpreting the data.

First of all, every question in the questionnaire will be computed into variables such as V1, V2, V3, V4 whereas data such as the levels in Likert scale will be numbered as 1,2,3,4, and 5. These data will be entered into the SPSS for statistical analysis.

Next, descriptive statistics will be used to interpret the data such as frequencies, measures of central tendency and dispersion. The histogram will be used to interpret the data showing the frequency, percentage, valid percentage, skewness of the graph, and also the cumulative percentage of the data. Furthermore, data will be analyzed through mean, mode, median for the measures of central tendencies. On the other hand, measures of dispersion provide data in the form such as standard deviations.

Cronbach's Alpha will be processed to test both the consistency and stability of the data obtained. And finally, The Pearson Correlation Analysis will be conducted to verify the afore-formulated research hypothesis, that is, the correlations between independent variables and dependent variables.

6. EXPECTED RESULTS

From this research, it is expected that the outcomes achieved will indicate whether the determinants have a positive or negative relationship with the purchasing intention of

Malaysian consumers towards green products. In the sense that if the results yield show a positive relationship between attitude variable and the intention variable, it will present a significant meaning that the tendency of Malaysian consumers purchase for green products depends on their personal attitudes. Meanwhile, if the outcomes obtained prove that there is a negative relationship between attitude variable and the intention variable, it will indicate that the purchasing intention for green products is not affected by their attitude.

7. SIGNIFICANCE OF STUDY

The findings of this study could facilitate the marketers as well as policy makers to formulate their policy and marketing plan which would promote the purchase and usage behavior of the consumers towards green products. It is beneficial for the policy makers to strive towards improving environment by arousing people to conserve the environment and ecology from their behavioral intention aspects of the consumption. This study would also provide valuable information for marketers in predicting Malaysian consumer's purchasing intention for green products, hence, to more efficiently promote their green products and services.

ACKNOWLEDGMENTS

The authors thank the Malaysian Technical University(MTUN) COE for Funding RDU 121214.

REFERENCES

1. Yale University., 2011, " Environmental Performance Index," retrieved from http://epi.yale.edu/, accessed on 1 April 2013.

2. The Star., 2012, " M'sia improves in Environmental Performance Index," retrieved from http://thestar.com.my/news/story.asp?file=/2012/1/27/nation/20120127175805&sec=nation , accessed on 7 April , 2013.

3. Ottman, J. A., Ed., 1998, "Green Marketing: Opportunity for Innovation," NTC Business Books.

4. Pavan Mishra, P. S., 2010, " Golden Rule of Green Marketing," Green Marketing In India: Emerging Opportunities and Challenges 3: 6.

5. Mainieri, T., E. Barnett, T. Valdero, J. Unipan, and S. Oskamp., 1997, " Green buying: The influence of environmental concern on consumer behavior," Journal of Social Psychology, 137(2), 189-204.

6. Kim, Y., and Choi, S.M., 2003, "Antecedents of pro-environmental behaviors: An examination of cultural values, self-efficacy, and environmental attitudes" International Communication Association, Marriott Hotel.

7. Our Green World., 2008, "How green is our world? Research Report, TNS," retrieved from http://www.wpp.com/~/media/SharedWPP/Marketing%20Insights/Hot%20Topics/Climate%20Change/TNS_Market_Research_Our_Green_World.pdf, accessed on 7 April , 2013.

8. Steg, L., and Vlek, C., 2009, " Encouraging pro-environmental behaviour: An integrative review and research agenda," Journal of Environmental Psychology, 29, 307–317.

9. Ajzen, I., & Fishbein, M., 1980, "Understanding Attitudes and Predicting Social Behavior," Englewood Cliffs, NJ: Prentice-Hall.

10. Lee, K., 2008, " Opportunities for green marketing: Young consumers," Market. Intell. Plann., 26: 573-586.

11. Boldero, J., 1995, " The prediction of household recycling of newspapers: The role of attitudes, intentions, and situational factors," Journal of Applied Social Psychology, 25 (5), 440-462.

12. Chan, K., 1998, " Mass communication and proenvironmental behavior: Waste recycling
in Hong Kong," Journal of Environmental Management, 52, 317-325.

13. Ajzen, I., 1991, " The theory of planned behavior," Organizational Behavior and Human Decision Processes, 50, 179-211.

14. Perner, L., 2013, " Consumer behavior: the psychology of marketing," retrieved from http://www.consumerpsychologist.com, accessed on 7 April, 2013.

15. Taylor, S., & Todd, P., 1995, " An integrated model of waste management behavior: A test of household recycling and composting intentions," Environment and Behavior, 27, 603–630.

16. Vermeir, I., & Verbeke, W., 2006, " Sustainable food consumption: exploring the consumer attitude-behavioural intention gap," Journal of Agricultural and Environmental Ethics, 19(2), 1-14.

17. Ferber, R., 1977, " Research by convenience," Journal of Consumer Research, 1, 57-58.

18. Do Valle, P.O., Reis, E., Menezes, J., and Rebelo, E., 2005, " Combining behavioral theories to predict recycling involvement," Environment and behavior, 37, 364-396.

19. Vermeir, I., and Verbeke, W., 2006, " Sustainable food consumption: exploring the consumer attitude-behavioural intention gap." Journal of Agricultural and Environmental Ethics, 19(2), 1-14.

20. Sparks, P., and Shepherd, R., 1992, " Self-Identity and the theory of planned behavior: Assesing the role of identification with "Green Consumerism"," Social Psychology Quarterly,55(4), 388-399.

21. Roberts, J. A., 1996, " Green consumers in the 1990s: Profile and implications for advertising," Journal of Business Research, 36(3), 217-231.

22. Baker, Michael J. and Gilbert Churchill A. Jr., 1977, " The Impact of Physically Attractive Models on Advertising Evaluations," Journal of Marketing Research, 14 (November), 538-555.

*****

Leadership and Characteristics of Women Principals In Secondary Schools In Johor Bahru

Shanthi Bavani V Raja Mohan, Ph.D.

Academic Editor: *Shahryar Sorooshian*

ABSTRACT: Effective school leadership is one of the significant key to implement reform policies and coodinate the principles accordingly. In addition, it also depends largely on the ability of the leaders to develop and engage in complex thinking, to analyse, solve problems and face challengers. This study aim to determine the women leadership styles and challenges in secondary schools in Malaysia. Six leadership dimensions are outlined namely school objectives and policies, curriculum and teaching management, monitoring and evaluation, support and motivation, staff development and lastly communication.

Keywords: School Principals, School Leadership Styles, Challengers

A. INTRODUCTION

Today's women leaders demonstrate strength, dignity and power to face challengers and build more diverse future in many aspects. The representation of women in leadership roles shows the level of intelligence, contribution and their commitment in achieving the objectives of an organization. This research on women principals from five (5) premium schools from Malaysia has been selected to understand their leadership skills, management and challenges faced and lessons learned in their respective schools that has contributing to the Malaysian society.

Women in school administrations and their relationships with others were central to all actions. Women principals communicated more, motivated more, and spent more time with teachers and students in most of their everyday life Morale was higher and relations with parents were more favourable. The perspective is of the morality of response and care, which emphasises maintaining relationships and promoting the welfare of others. Teaching and learning are the main focus. A proper school climate should be developed to emphasise instructional programs and enhance student progress.

The women's characteristics are democratic, participatory and encourage inclusiveness besides a broad view of the curriculum. Women principals boast leadership characteristics that are conducive in promoting good schooling. They have clear educational goals supported by a value system that stresses service, care and relationships; they are focus on instructional and educational issues and build a supportive atmosphere; and they monitor, intervene and evaluate.

This study is about women characteristics as secondary school principals. I admire tremendously how women teach and lead. The method of teaching is inclusive, and is lead by consensus which recognises the value participants bring to the group. Leaders are aware that with authority comes responsibility and influence should be used wisely and sparingly. Leadership is shared.

Leadership is talked about all the time, but can only be achieved in an environment that promotes this kind of mutual trust and acceptance of individual rights, beliefs, and responsibilities. Respect is another important theme threading the discussions throughout the interviews by the participants. I see our society changing very quickly and civility and respect are often not valued in our busy lives. Since I am involved with students, it is hard to separate myself from my job. Students are my passion. I enjoy helping students put things into perspective. I know the importance of being persistent because the time it took for me to complete my Degree and the dropouts, or stop-outs on my journey through life and school is a clear testimony.

Signals and established expectations such as these stifle the more positive qualities that women are socialised to develop. Hence, this must be studied for women to consistently achieve success in educational leadership positions. Women who aspire to be school administrators must identify specific factors that impede their professional growth and must develop skills to surmount these obstacles. The other aspect to be studied is to determine the leadership characteristics that women principals bring to the position. Many research studies have been conducted to determine what makes leaders successful. Leadership may be defined as the ability to get people to work together to accomplish common goals. Numerous characteristics such as intelligence, charisma, integrity, attitudes and people skills have been identified as skills needed to become a successful leader.

B. DEFINING LEADERSHIP

Leadership may be defined as the ability to get people to work together to accomplish common goals. Koontz (Zaidatol, 1993) defined leadership as a skill to influence staff and employees to complete their tasks with great interest and confidence. . According to Leithwood and Reihl (2003), in the field of education, as in many other institutional contexts, leadership has taken on increased importance in recent years. Leadership is also a process of a principal to obtain synergy and cooperation from his or her staff in order to realise a certain objective and also a process when the leader of a group manages to steer the group to fulfill an objective (Razali Mat Zain, 1996).

According to Zaidatol, (1993), leadership is a process to direct employees' efforts towards achieving organisational mission. Furthermore, Daft (2005) argued that leadership is about "creating a compelling vision of the future and developing farsighted strategies for producing the changes needed to achieve that vision". In summary, it can be seen that leadership is fundamentally situated in both context and relationship.

C. LEADERSHIP QUALITIES

Qualities are defined as a character, behaviour or habit that belongs to someone summarises that the qualities of a leader encompass the ability to handle one before leading others, ability to manage, boast communication concepts and skills, honest and trustworthy plus diligent. Conger and Kanungo (1988) states that leadership qualities consist of strong commitment towards a goal, believes in oneself and is an agent for radical changes, and not a manager from status quo. Quoting Islamic tenets, a great leader must have qualities including highly intellectual, trust in oneself, ability to negotiate and discuss, highly motivated and full of initiatives and aspirations. Women leaders that are relationship oriented will surely receive a great commitment from their employees. Task oriented women leaders will surely received objections from their employees because they are forced to work without any open communications.

D. QUALITIES OF A GOOD PRINCIPAL

Hallinger (1987) did a research in schools all over Australia and discovered that excellent schools boast principals with the qualities below:

i. Responsible and supports teachers' needs

ii. Realise and is aware of happenings in classrooms

iii. Enables share of responsibility and resources in the most effective way

iv. Flexible leadership characteristics

v. Programs are constantly evaluated and updated

vi. Always provide positive feedback

vii. Prepare resources to fulfill school requirements

viii. Maintain a good relation with the Education Department, community, parents, teachers and students

ix. Ever ready to take the risk

x. Concern about personal profesional enrichment and constant development

xi. High level of commitment towards mission and values of teachers and students

xii. Establish good moral values like respecting others among students

Principal is the key to paradigm shifts towards an effective school. This will then be a platform for all academic and non-academic staff to create an effective school. According to Mortimore et. Al (1988), the profesional leadership of a principal involves experience and knowledge related to classroom handling, curriculum planning and execution on top of efficient teaching strategies. Sizer et. al (1984) added further on the principal's role as the leader of projects, source of encouragement and often visits classrooms and have informal discussions with staffs. Resulting from this excellent leadership, there will be a positive quantum leap in the students' performance. This will then influence the school's development from the work culture perspective, classroom effectiveness and quality of teaching.

Norasimah Rahim (1990) discovered that excellent secondary schools in Kelantan, Malaysia boast the following recipe of success:

i. Principal plans all school activities seamlessly and effectively
ii. Principal includes teachers in decision making
iii. Principal monitors and controls the implementation and execution of school activities
iv. Teachers enjoy working and work satisfaction is achieved
v. Discipline management is effective
vi. Schools' relation with parents and society is maintain at a good level
vii. Implement special programs for schools' excellence in all areas
viii. Delegation of work is clear and acceptable
ix. Principal – teacher communication is at an excellent level
x. Principal evaluates every activity planned and take measures to counter problems if any
xi. Principal convey schools objectives and vision to all teachers

Hanson (1979) elaborated that transactional leadership characteristics take into consideration various requirements depending on the task situation. These characteristics of leading also assume that different problems require different leadership characteristics. Quantz (1991) equates these transactional characteristics to the barter system. Burns (1978) put forward a different theory called transformational leadership characteristics, which is of course entirely different from the transactional characteristics. Transformational leaders are great role models to their staff. Nik Aziz Nik Pa (1991) wrote on the leadership stages from the Islamic perspective. He noted that these various stages are dominant factors to change and strengthens leaders. The leadership hierarchy are slow leadership, intelectual leadership, professional leadership and ideal leadership.

Whatever the approach and qualities of the principals, it must be suitable with the school he is leading. There is no one approach for all situations. It is up to the principal's wisdom to balance all factors and variables that will shape the school. Every leader has their own set of characteristics and qualities and everyone is different from one another. Generally, a leader must be stern, agressive, independent, progressive and visionary. Task oriented leaders have full control over its subordinates. People oriented leaders stressed more on democracy and cooperation (Ahmad Atory, 1996). Hence, principals as leaders of schools must adapt to situations and apply different qualities to each set of challenge or problems. The principal must understand his or her staff in terms of work ethics and methods, personal matters, motivation levels, objectives and level of satisfaction for each staff.

E. LEADING THE SCHOOL

Schools are the thrust of our National Education. Human resource development begins with the education system and in schools. In our quest to be a learned society, schools play a significant role. Principals leading this ark to success therefore hold a pivotal position and are in fact the catalyst to this dream. The achievement and excellence of a school is closely related to the leadership of its principal (Abdul Shukor, 1998). In this new age, school management and leading is not as easy as once thought. The principal will now need to master many fields including education, technology, management, economy and leadership ways. According to Mazlan Abdullah (1997), the essence to success in all school principals lies in their positive attitudes, vast knowledge, vision and mision, management, human relations and communication skills. Principals should have the edge in terms of mind capacity, skills and knowledge, vision and strategies for implementation. All these are in efforts to produce a new generation that is capable to continue our nation building.

Women principals are not a new addition to the education system and have started decades ago. Although the numbers are small, it surely is increasing as more women are involved in education (Zaneriah, 1998). A principal who successfully led a school will be able to show continuous excellence and overall achievement in academic curriculum, co-corriculum, teachers, students and staff. A well lead school will outperform itself and went beyond its own expectation. An effective school must encompass not only efficiency in the institutional level but must relate this effectiveness to all levels of government including nation's development with noted emphasis on human resource development (Abdul Shukor, 1998). The effectiveness of schools can be benchmarked in the district, city, state and nationwide levels.

F. SCHOOL EVALUATION CHANGES

Schools are getting bigger and the evolution of the education system four decades ago spurred the constant growing number of students. These factors eventually made it a lot more challenging to manage a school in Malaysia (Al Ramaiah, 1992). It is a well known fact that schools in Malaysia are not only different from the cultural aspects, but also different when it comes to leadership characteristics and management practices (Zaneriah, 1999). There are autocratic principals as well as democratic ones.

Experience does help us face the challenges in lives. New principals have to rely on their wealth of knowledge to tackle the tasks at hand or refer to other more experienced principals. What it takes for women to lead a secondary school as its principal is not well known. Hence, this study will determine the leadership characteristics and qualities required to be a women principal in a secondary school in Malaysia. A multitude of challenges become stumbling blocks for women principals. We need to study the various barriers and challenges facing an aspiring woman in her quest as a secondary school principal.

G. IMPORTANCE OF THE STUDY

This research can benefit many parties. The results of this research can help principals to re-evaluate their roles as managers and leaders of schools. It can also be an encouragement for women principals to do things right, remedy the problem and overcome their weaknesses and shortcomings in leading their schools. Principals can take this research as a guideline to various concerted strategies for a better and more robust leadership. This study can also determine the barriers for women to assume roles as principals. This can surely assist them to prepare themselves well in order to realise their aspirations. The Ministry of Education can also use this research as a basis to determine more consistent criteria for selecting school principals. Moreover, the feedback from this study can be utilised by institutes of higher learning to train undergraduates in management and leadership programs.

E CHARACTERISTICS OF WOMEN LEADERSHIP

A characteristic is defined as a distinguishing trait, quality or property (Oxford Dictionary 2004). Qualities is defined as a character, behaviour or habit that belongs to someone (Oxford

Dictionary, 2004). Ahmad Sahiri (2004) summarises that the qualities of a leader encompass the ability to handle oneself before leading others, ability to manage, boast communication concepts and skills, honest and trustworthy plus dilligent. Conger and Kanungo (1988) states that leadership qualities consist of strong commitment towards a goal, believes in oneself and is an agent for radical changes, and not a manager from status quo. Quoting Islamic tenets, a great leader must have qualities including highly intellectual, trust in oneself, ability to negotiate and discuss, highly motivated and full of initiatives and aspirations. Task oriented women leaders will surely received objections from their employees because they are forced to work without any open communications.

F.　PAST RESEARCHES LOCALLY

Researches done on leadership of principals in Malaysia are broad indeed. A few local researchers gave their opinion on this topic. Ahmad Sahiri (2004) observed female leadership qualities and its impace to their workers' commitment. Results showed that women leadership from the perspective of custom officials are at a medium level. Secondly, their commitment towards a woman leader is also at a medium level. Abdul Karim (2001) used the instruments adopted from Leithwood et. Al (1992) to reveal that women principals give their most attention towards achieving goals. Mukerjee (1970) found that relationship between principals and teachers is also one of the main factors to define their respective responsibilities. Therefore, teachers must be clear of their role to produce a high degree of effectiveness in their work.

According to Zaneriah (1998), there is no significant difference between daily secondary schools and religious Arab schools when it comes to principal management. Although this is so, there is an existence of conflicting perception by teachers and students on the management techniques brought up by principals. Stoner (1984), found that there is a normal connection between the teachers' values and the management characteristics of a principal. Alageswary (1984) uses the LBDQ instrument to observe headmasters' leadership in primary schools in Selangor and Kuala Lumpur, Malaysia. The results outlined that most headmasters there are task oriented.

Abdul Hamid (1997) conducted a research to examine the perception of teachers and staff towards the principal's management in the district of Johor Bahru and Pontian, Johor, Malaysia. He showed that there is no difference in perceptions based on gender when dealing with leadership of the principal. Yin Cheong Cheng (1997) did a research on the perception of leadership among women principals and the teachers' work behaviour. For this research, leadership among women principals are divided into five dimensions, humanity, structure, politic, symbolic and education. Teachers' attitudes are classified into three sections that is work satisfaction, approach towards work and a leader plus their affection towards the profession as a teacher. This study suggested that changing perception is pivotal in efforts to revamp teachers' approach towards the profession. The women principals must take a step forward and diversify their leadership characteristics before they are able to diversify the teacher's perception.

G. PAST RESEARCHES ABROAD

Alimo-Melcalfe (1998) did a research on how the gender of a person is irrelevant as a manager. The difference between men and women in the world of management was observed. It was discovered that women give more attention to tasks in comparison to men. It is also shown that women is greater in the aspect of team building and to shape up workers. Women managers are more innovative, a better communicator, manage activities well and could act as mentors to their employees. James S. Pounder (2002) did a detailed research to observe who are the better leaders in management and education – men or women. Five (5) categories for leadership were detailed – structure, human, political, symbolic and educational. Patricia Hind (1997) did a perception research to look at gender differences in the value system and profession building. Several organisations were selected and must have operated for 5 years, boast high technology and have at least 100 employees. Results outlined that women are holding lower positions and receive less salary as compared to their male counterparts. From previous researches and studies, be it local or from abroad - women leadership is pivotal and contributes immensely to the growth of information and intellectuality.

H. SUMMARY

Each responsibility should be executed well and accordingly. A leader must never put fame or name as his or her priority. Rewards in the form of salary, incentive and bonus are signs for a job well done. Leadership in education covers all activities, actions, skills and attitudes, inclusive of knowledge and character, ability to convince, move, train and achieve the school's objectives and how to set quality standards and benchmarks; and how to manage monitoring between the school-site and the system level. An effective leader is one who can define freedom that should be awarded to their workers and the correct time to do so. To enhance efficiency and productivity, principals should find an initiative theoretically or practically to enhance their management and leadership system.

REFERENCE

Abdul Shukor Abdullah (1998). "Fokus Pengurusan Pendidikan". Kementerian Pendidikan Malaysia

Ahmad Atory Hussain (1996). Pengurusan Organisasi, Kuala Lumpur. Utusan Publications & Distributors Sdn. Bhd.

Alimo-Melcafe (1998). Dipetik dalam Tricia Vilkinas (2000). "The Gender Factor in Management: Women in Management Review, Volume 15 Number 5/6. 1 -14

Al-Ramaiah (1992). Kepimpinan Pendidikan: Cabaran Masa Kini, Petaling Jaya. IBS Buku Sdn. Bhd.

Burns, J.M. (1978). Leadership. New York: Harper and Row Publisher

Conger, J.A and Kanungo, R.N. (1998) Charismatic Leadership – Josey Bass Publishers

Daft, R.L. (2005). The leadership experience (3rd edn). Mason, Ohio: Thompson South-Western.

Hallinger. P., (1987). Institutional Leadership in School Context, in W. Greenfield. Boston: Allyn and Bacon

Hanson, M.E. (1979). Educational Administration and Organisational Behaviour. Massachusetts: Allyn and Bacon

James S. Pounder, Marianne Coleman (2002). Women Better Leaders than Men? Leadership Organisation Development Journal, Volume 23, Number 23

Leithwood, K. et. al (1992). Developing Expert Leadership for Future Schools: New York

Mazlan Abdullah (1997). Kepimpinan Pengetua Dalam Pengurusan Staf Sokongan- Satu Kajian Kes. Dalam Seminar Nasional Pengurusan.

Mortimore, P. dan Mortimore, J. (1988). Teachers Appraisal: Back to the Future. School Organisation. Vol. 11. No 2.

Nik Aziz Nik Pa (1991). Kepimpinan dan Pengurusan Sekolah pada Zaman Maklumat, Jurnal Guru.

Norasimah, R. (1990). Ciri-ciri Pengurusan dan Kecemerlangan Akademik di Sekolah Menengah Kelantan. Universiti Teknologi Malaysia.

Quantaz, R.A., Roger, J. Dantley, M. (1991). Rethinking transformative leadership: Towards Reforms of Schools. Journal of Education.

Razali Mat Zin. (1996). Kepimpinan dalam Pengurusan. Kuala Lumpur. Utusan Publications and Distributors.

Sizer (1984). Horace Compromise:The Dilemma of the American School. New York: Hoghton Mifflin Company

Stoner (1984). Management. New Jersey: Prentice Hall Inc.

Zaidatol Akmaliah Lope Pihie (1993). Pentadbiran Pendidikan. Shah Alam: Fajar Bakti

Zaneriah. M (1998). Amalan Pengurusan Pengetua ke Arah Kecemerlanagan Akademik Sekolah Menengah. Universiti Teknologi Malaysia.

*****

A Comparison between Green Concrete and Conventional Concrete

Assef Shakeri

UPM University, Serdang, Malaysia

Academic Editor: *Amin Teyfouri*

ABSTRACT: Concrete is the most popular construction material worldwide. Its popularity is for many factors, such as availability, low cost, vast applicability in all fields of civil. But concrete also involves a high environmental cost. The billions of tons of extracted and processed natural materials annually are intended to create significant problems on the environment. The most problem is the huge amount of energy needed to produce Portland cement. Also large amounts of CO_2 emitted into the atmosphere in the cement manufacturing process.

This paper summarizes the various initiatives to improve the environmental performance of concrete to make suitable material, such as "Green Concrete". By using appropriate substitute materials instead of cement such as fly ash, silica fume and Rice hull ash. As well as it attempts to use the proper recycled materials as concrete aggregate replacement are very important in this issue such as recycled concrete aggregate, junk or crushed glass, rubber, etc.

A large number of buildings around the world were constructed in the Fifties and Sixties therefore they are close to end of their life. it can be predicted that during the next decades, there are many demolished concrete, as a result of the destroying of these buildings and structures, will be available for recycled concrete aggregate and able to be reuse for concrete. For improving sustainable building materials high volume fly ash mixes for various civil structures were introduced and developed. Other existing methods like self-compacting concrete or ICF forms were developed for improve economics and sustainability.

In this study, three primary mixes. type1, type2, type3 were designed for comparison green concrete with conventional concrete. Two types were made by recyclable aggregate and 10-30 percent supplementary cementitious material (rice hull ash) another type was mad by natural aggregate. These mixes were tested in civil laboratory at UPM University for a number of factors such as compressive strength, set time and workability.

The study examines some of the economic factors that show the degree of trade success. Through recognition and utilization of intrinsic properties in difference waste materials or products, which can improve value of these materials and can achieve successfully in a market-oriented economy of supply and demand.

Keywords: concrete; sustainability; green building; recyclable aggregate; supplementary cementitious materials; green concrete

INTRODUCTION

Green building materials so that is clear from its name, they are construction materials that are environmentally friendly by reducing maintenance cost and energy, improving occupant health and productivity, lower costs related to changing space configurations Building and construction activities worldwide consume 3 billion tons of raw materials each year or 40% of total global use [1]. Using green building materials and products, will promote protection of declining nonrenewable resources internationally. In addition, integrating green building materials into building and civil projects can help to reduce the environmental impacts associated with the extraction, transport, processing, remanufacturing, installation, reuse, recycling, and waste disposal of these building industry source materials.

Concrete is the most widely common material used that has wide application in building and civil structures worldwide. Concrete is popular because, it is durable, low cost and it is available everywhere (general availability) [2]. Today, concrete industry has been developed and concretes with various properties are on the market for example, fibrous concrete with more tensile and flexural strength conventional concrete which is made by inorganic polymers such as alumina-silicate (Al-Si).

Concrete structures are superior compared with steel structures based on the following reasons [3][4][5]:

- concrete plasticity is more and better than steel
- Raw materials of concrete can be found in most parts of the world
- Concrete structures have good resistance against fire
- Compressive strength of concrete is great(the only weakness of concrete is low tensile strength)

Although, in terms of impact on the environment is one of the most costly material. Conventional concretes are composed of 80%aggregate (coarse and fine) and 12%ciment.

In terms of concrete volume,[7]

- Cement: about 7% to 15%volume of concrete
- Water: 14% to 21% of concrete volume
- Aggregate: 60% to75%of concrete volume
- Air:

a) Concrete without air, air volume from 0.5 to3 percent

b) Concrete with air, air volume from 4 to 8 percent

- Additive materials: small amount of them add to concrete to provide desirable properties

The energy that is consumed to produce one ton cement is 4 giga-joules. Cement production generates large amount of CO_2 annually about 7%. In addition, excessive resources extraction such as limestone, clay, fuel and coal for creation concrete leads to make environmental problems like denudation, erosion, and top soil loss. 5% of total world greenhouse gases are caused by cement manufactures [8]. The lower amount of cement in concrete and use of other materials instead of cement in concrete mixture (such as supplementary cementation materials) will produce less CO_2 [9]. The Billions of tons of natural materials mined and processed each year, by their sheer volume, are bound to leave a substantial mark on the environment.

$Caco_3 + o_2 + heat\ (1200\text{-}1400^{oc}) = cao + co_2$

Effective use of concrete and cement will result in more stable structures especially for buildings and pavements.

The data for this paper were collected from the literature and laboratory results [10].

In this study was used of opinion of mainly architects, engineers, contractors and builders, developers, and consultants who have a strong interest or involvement in the field of sustainable building. A search was conducted to find existing reports on building products and their role in the design and construction of green buildings.

GREEN CONCRETE

Concrete is a building material which is inherently an environmentally friendly material, it must be used as much but with as little Portland cement as possible to decrease some challenges in construction like co2.the ways to increase concrete compliance with the demands of sustainability are [11][12]:

- Use of more supplementary cement materials such as fly ash, it leads to less use of cement in concrete
- Increased use of recyclable materials, it will reduce the demand for virgin materials.
- Improved durability and mechanical properties.

Making conventional concrete will damage to environment in various ways, one of them is natural resource extraction. some of negative impacts of extraction and use of natural aggregate: the effect on the natural habitats of many organism, also effect on the quality of water and increase greenhouses gas emissions that is caused by different processes such as mining, transport operation.

The purpose of this study is to develop a concrete with recycled materials, as well as using less cement and making concrete produced using environmentally friendly materials. The concrete's advantages than conventional concrete are that would create less CO_2, extraction resources and waste with lower energy consumption, spend lower cost and no pollution to environment[13].

AGGREGATE

Concrete structure must be demolished after useful life which provides a huge amount of waste. Two of the characteristics of aggregate particles that affect the properties of concrete which are particle shape and surface texture. A comprehensive study of the comparison GC and normal concrete (concrete made by natural aggregate) is presented in this paper. The main goal of the study is,

- Decreasing CO_2 and energy consumption by using supplementary cementitious material such as fly ash instead of cement.
- Decreasing demand for virgin material by replacing recyclable aggregate in concrete mix.

- Creating a concrete with low cost which has properties of typical concrete.

There are three classes specific gravity for concrete:

a) Typical concrete, the specific gravity for typical concrete is 2200 to 2500 kg/m^3

b) Lightweight concrete is a type of concrete that is made by lightweight aggregate such as volcanic pumice. Its SG is 1/2 to 1/3 SG of typical concrete. This concrete is used more for building facade, separating walls and dropped ceiling.

c) Heavyweight concrete, in this type of concrete is used steel and iron filings instead of sand.it can prevent of x-ray and γ-ray radiation and often used for the construction of nuclear facilities. Its SG is 1.5 to 2.5 SG of typical concrete (3500 to 6000 kg/m^3) [14].

In this paper compressive (f'_c) and tensile (f'_t) strength in 28 days were computed for 3 various concrete. Other properties of concrete, such as elastic modulus, water tightness or impermeability, and resistance to weathering agents including aggressive waters, are directly related to strength and can therefore be calculated from the strength data. After comparison strength, the amounts of their workability were compared.

Specific gravity is significant in important construction such as dams it is not normally considered however, low specific gravity will increase stability.SG of RCA is less of NA specific gravity. Water absorption as well as elongation and flakiness affect workability.

Elongated and flak particles, they are indicators for concrete workability. This test is used to specify the amount of aggregated particles that are elongated shape instead cubicle. It can be significant in some applications such as concrete, in which an elongated shape has a larger surface area and therefore require large amounts of reinforced concrete to produce the necessary strength. Flakiness index is the percentage by weight of particles in it; at least the size (i.e. thickness) is less than three-fifths of their average size [15]. Elongation ratio is the weight percentage of particles in it, with the longest side (i.e., the length) is greater than one and the average size of four-fifths times.

Flakiness index is measured by weighting aggregate passing via thickness gauge.

Flakiness index=Flakiness (flakiness between 30% to40% is appropriate for concrete)

[16]

Elongation index=*100   [17]

The tests will be done for aggregate with under dimensions of 6.3_{mm}. Amount of f and e for masonry concrete aggregate is more than demolished concrete. E and f for RCA is near to granite aggregate.

Major different between RCA and NA is residual mortar attached to the former. The residual mortar leads to lower specific gravity and lighter absorption capacity, beside the RCA is more porous than NA, it can absorb more water resulting it has better adhesion.

Concrete porosity must be low due to high porosity causes to decrease the strength, durable and leads to increase probability of freeze and thaw resistance of concrete.

Porosity for NA is 0 to 5 percent and for RCA is range of 12 to 50 percent compared to 6% of NA.

Generally, recyclable aggregate are more porosity, rough, coarse and grain sizes are closer together and are more fragile than natural aggregate.

Physical properties consists of, absorption capacity, density, porosity and permeability.

Mechanical properties are; compressive strength, tensile strength, flexural strength and elastic modulus [19].

Mechanical properties of concrete constructed by RCA are:

- sum strength with original concrete
- percentage of coarse to fine aggregate used in the RCA concrete is same with original concrete
- RCA concrete eliminates abrasion and increases water absorption.

Usually, coarse aggregate occupy about 75% of total volume of concrete. Although their properties may vary according to source, as well as, RCA manufactured and amount of residual mortar attached .generally mortar in RCA considerably affect its properties in fresh and hardened concrete, for example mortar attached to RCA improves mechanical and durability of RCA properties but it decreases specific gravity and makes higher absorption for concrete. In this study aggregate is supported from two batch, one them is recyclable aggregate of UPM university civil engineering lab, the other is a concrete plant in Isfahan, IRAN. RCA is from demolished concrete in both batches. RCA is obtained by mechanical stress (in UPM lab RCA obtained by hammer blows) or chemical stress (in Iran by sodium sulfate).

Observations have shown that, 30% RCA coarse aggregate and 20%fine aggregate can be used in concrete without reducing the strength but more than this amount will reduce strength

CEMENTING MATERIAL (RICE HULL ASH)

Recently, cementing materials are a main key in concrete admixtures. In many countries various types of fly ash are more common to use in concrete mix, but fly ash in some places doesn't arise. For example in middle east fly ash is less or is more expensive than cement while in north of America or Canada exists many fly ash with low price. Therefore in this paper rice hull ash has been applied for making concrete. Rice hull ash almost exists in all countries; Rice Hull Ash (RHA) is a pozzolanic material wrought by burning rice hulls and straw, usually it is associated with generating electricity. RHA has several properties similar to silica fume. It is quite similar in chemical composition.

Some equation needed for Concrete mix design

In concrete mix design three issues must be considered:
- achievement to the desired strength

- providing sufficient durability
- achievement to desired slump

In this study to calculate the compressive and tensile strength, some equations were used:

Equation to compute moisture content (MC):

MC (OD) =   [20]

W_S: is weight of aggregate in stock

W_{OD}: is weight of oven dry aggregate

Equation to moisture content SSD:

MC (SSD) =   [21]

W_{SSD}: is weight of SSD aggregate

Equation of absorption: [22]

In designing concrete mix; SD moisture is used as criterion because that is a balance condition which aggregate neither absorb water not give up water to the mix.

- Density and specify gravity

Density (D): basically density has been defined the weight per unit volume:

D =   [23]

V_S, is volume of solid

Bulk density (BD): pores volume inside a single aggregate

BD =   [24]

Specify gravity (SG): the ratio of mass of a given substance divided by unit mas of an equal volume of water.

Absolute specific gravity (ASG) =   ,   w is unit mass of equal volume

Bulk specific gravity (BSG) =   [25]

In concrete mix BSG value is used usually, BSG can be recognized by Archimedes' principle. Weight of sample is first measured in air (W_{SSD} in air) after that in water (W_{SSD}). But BSG for most rocks is 2.5 to 2.8.

BSG_{SSD}=   [26]

Experiments show that tensile strength is 10 to 15percent of compressive strength.

f'_t= (10% to 15%) f'_c

Unit weight (UW):

UW=   [27]

And, percentage of spacing is:

Spacing (void) =   [28]

- sieve analysis:

Sieve analysis is done to estimate grain size as well as is a way to separate fine and coarse aggregate. Sieve analysis determines the proportion of different size of particle and records them. This record is the result of analysis.

Percent passing of aggregate = 100   [29]

W_t=total weight of aggregate

W_s= retained aggregate on each sieve

Grain passed from sieve No.50 must be less than 10% to 30% of total weight of fine aggregate [30]. Too much fine in mix leads to increase amount of cement however low fine aggregate in mix raise amount of water. In addition it will decrease concrete workability.

In this study recyclable materials were recyclable concrete aggregate while there are a huge numbers of waste materials that can be applied in concrete mix. For example, many interior designers utilize from concrete that is made by waste glass with different colors for designing offices or restaurants or even homes instead of using wood, paper, etc [31][32].

Three sources of materials used in this study where the first source was, demolished concrete which was made for various tests in UPM University civil lab and after test was discarded(type1).second source was demolished concrete of a grouting and concrete plant in Iran (type 2)and the last source was Malaysia natural aggregate(type3). The type 1 is concrete made by RCA aggregate obtained from Malaysia with 10%-30% supplementary cementitious materials so that they replaced instead of cement type 1 in mixture [33]. Type 2 is similar to type 1 but RCA used in concrete type 2 is, demolished concrete of Iran. Finally, type 3 is concrete made by virgin aggregate without cementitious materials.

6.2 MIX DESIGN:

In this paper to mix design ACI method is applied. Before mix design some parameters must be computed:

- RMC of the RCA
- Specific gravity and absorption capacity of the NA and RCA.
- Unit weight
- Fineness modulus
- Specific gravity of cement

The main steps for proportioning the constituents are:

- Estimate w/c
- Calculation the amount of cement
- Calculation of aggregate content

For comparison of RCA concrete with NA concrete must be prepared identical samples from both of them.it means that the weight, amount of cement and slump of both samples have to be equal [34].

Absorption capacity, porosity and permeability in concrete made by RCA are more than concrete made with natural aggregate.

The amount of slump for three samples was 50-100; therefore, other calculations were done according to slump. Given that the ACI tables, the results of concrete mix design type 1, type 2 and type 3 are calculated. Results in the tables are listed below:

Table 1: type1, RCA concrete (Malaysia)

Type 1	slump	Water-cement ratio	Coarse aggregate (kg)	Fine aggregate (kg)	Amount of cementitious materials (kg)	Fineness modulus	Maximum size of aggregate (mm)
Sample 1	50	0.788	11.81	7.61	0.268	2.65	25
Sample 2	75	0.786	11.80	7.49	0.275	2.65	25
sample 3	100	0.79	11.80	7.50	0.275	2.65	25

Table2: type 2, RCA concrete (Iran)

Type 2	slump	Water-cement ratio	Coarse aggregate (kg)	Fine aggregate (kg)	Amount of cementitious materials	Fineness modulus	Maximum size of aggregate (mm)
Sample 1	50	0.73	11.16	8.51	0.260	3	25
Sample 2	75	0.74	11.16	8.26	0.275	3	25
sample 3	100	0.74	11.16	8.48	0.262	3	25

Table3: type 3, natural aggregate (NA) concrete

Type 3	slump	Water-cement ratio	Coarse aggregate (kg)	Fine aggregate (kg)	Amount of cementitious materials	Fineness modulus	Maximum size of aggregate (mm)
Sample 1	50	0.695	10.27	8.80	0	3.1	20
Sample 2	75	0.7	10.27	8.78	0	3.1	20
sample 3	100	0.7	10.27	8.41	0	3.1	20

Type	$W_{SSD(coarse\ aggregate)}$ %	$W_{SSD(fine\ aggregate)}$ %
Type 1	0.8	0.7
Type 2	0.7	0.7
Type 3	0.5	0.7

Table 4: compressive strength at 28 days for concrete type3

Type 3	compressive strength at 28 days (MPa)	Weight(kg)
Sample 1	16.668	2.342
Sample 2	16.608	2.324
Sample 3	16.576	2.312

Table 5: compressive strength at 28 days for concrete type2

Type 2	compressive strength at 28 days (MPa)	Weight(kg)
Sample 1	16.83	2.339
Sample 2	17.53	2.305
Sample 3	17.09	2.295

Table 6: compressive strength at 28 days for concrete type1

Type 1	compressive strength at 28 days (MPa)	Weight(kg)
Sample 1	18.63	2.241
Sample 2	16.95	2.232
Sample 3	16.76	2.220

Generally, if 100%coarse aggregate and 50% fine are used for concrete mix, compressive strength of concrete would decrease 20% to 30 % rather than NA concrete. Also its compressive strength 15% to 20% will be smaller than compressive strength of concrete made by combining NA and RCA.as the tables show, amounts of compressive strength in 3 types of concrete have insignificant difference. moreover the cost for conventional concrete is 3 times more than RCA concrete for example the price for making 1m³ typical concrete in Malaysia is approximately 135.1 Ringgit while this price for concrete made by RCA is about 48 ringgit(the rules of Malaysia.www.docstoc.com)the prices are without labor cost. Usually, compressive and tensile strength in the first 3 days for concrete made by RCA is more than NA concrete but in 7 or 28 days strength, compressive and tensile strength is more for concrete which made by natural aggregate.it can be because of good bonding between RMA and cement.

CONCLUSION

Economic of recycling is widely dependent on their application. All in all; natural (virgin) materials in terms of quality control have advantage than recyclable aggregate. However, over time, the economic possibility of recyclable material will increase, as well as natural aggregate will be scarce dramatically in addition, the excretion cost construction waste and other debris aggregate for destroyed construction continue to rise. Increasing green building principles and sustainability procedures will change the economic outlook for the support of the environment. We cannot continue to lose our natural resources. Governments are responsible for keeping the principles. They must support the cost of reuse, recyclable or decreasing waste landfilling.in many countries especially European countries, it is now a rule that manufactures and producers before designing and manufacturing a product must consider its waste cost.

Political conflicts are shown when you need to balance the requirements of environmental protection against the development of elevating the standard of living [35].it is clear that as we received the world from previous generation must pass on to next generation. While developed and industrialized countries are responsible to reduce environmental problems especially pollution and their participation of using resources of world such as energy, developing countries must not repeat past mistakes. It is predicted that, the production of cement and fly ash is rising particularly in Asian countries.in India and China, cementitious materials production will be increased considerably in next decade.

Improvements in last concrete research have shown many developments in making concrete by using RCA and supplementary cementitious materials that lead to minimize the need for cement manufacturing which can be a guideline to help sustainable development in the world.

One of the most important industries in using large amount of energy and virgin or natural resources is concrete industry. Also this industry generates a vast amount of CO_2 into the atmosphere annually. Applied recyclable material and substitute Portland cement can improve the concrete industry in next few decades. Now, we should work together to have a planet habitable.

This research presented a current state of green concrete and concrete investigation needs toward sustainability. The industry's suggestions or feedback is worth to correct the research problems of sustainable concrete in the academic world and create researchers with ideas to show the industry worries the use of sustainable concrete.

Table7:type1 grading aggregate

Sieve#	Retained%	Passing%
1"(25$_{mm}$)	28	72
3/4"(19.5$_{mm}$)	44	56
3/8" (9.5$_{mm}$)	58	42
#4(4.75$_{mm}$)	65	35
#8(2.38$_{mm}$)	77	23
#16(1.18$_{mm}$)	85	15
#30(600$_{\mu m}$)	90	10
#50 (300$_{\mu m}$)	95	5
#100(150$_{\mu m}$)	97	3
#200(75$_{\mu m}$)	98	2
>4.75mm% =	66.2	
4.75-0.075mm% =	33.6	
<0.075mm% =	0.2	

Table 8: type2 grading aggregate

Sieve#	Retained%	Passing%
1"(25$_{mm}$)	3	97

Sieve#	Retained%	Passing%
3/4" (19.5$_{mm}$)	8	92
3/8" (9.5$_{mm}$)	46	54
#4(4.75$_{mm}$)	70	30
#8(2.38$_{mm}$)	85	15
#16(1.18$_{mm}$)	93	7
#30(600$_{\mu m}$)	96	4
#50 (300$_{\mu m}$)	98	2
#100 (150$_{\mu m}$)	99	1
#200(75$_{\mu m}$)	100	0

>4.75mm% =	70.8
4.75-0.075mm% =	29.2
<0.075mm% =	0.1

Table9: type3 grading aggregate

Sieve	Retained	Passing
1"(25$_{mm}$)	0	100
3/4"(19.5$_{mm}$)	12	88
3/8" (9.5$_{mm}$)	47	53
#4(4.75$_{mm}$)	66	34
#8(2.38$_{mm}$)	83	17
#16(1.18$_{mm}$)	91	9
#30(600$_{\mu m}$)	93	7
#50 (300$_{\mu m}$)	96	4
#100(150$_{\mu m}$)	99	1
#200(75$_{\mu m}$)	100	0

>4.75mm% =	66.2
4.75-0.075mm% =	33.7
<0.075mm% =	0.1

REFERENCES

[1]. Concrete Admixtures Handbook, 2nd Ed.: Properties, Science and Technology V.S. Ramachandran

[2]. http://www.naiop.org/foundation/greenincentives.pdf

[3]. Emerald Architecture case studies in Green Building, Green Source Magazine, Chapter 7,Science/Technology, McGraw-Hill Companies , Inc., 2008

[4]. ProQuest Documents: Concrete as a Green Building Material C. Meyer Columbia University, New York, NY 10027, USA

[5]. How LEED Certification Can Pay Off for Multnomah County Portland state university

[6]. Element Analysis of the Green Building Process Richard E. Zigenfus

[7]. Environmental Building News. (1996) E-Build, Inc., Brattleboro, VT; 802-257-7300; info@ebuild.com http://www.ebuild.com

[8]. U. S. Green Building Council, project profile facts heet,usgbc.org/DisplayPage.aspx?cmsPageID=1721

[9]. www.allenmatkins.com/emails/GreenSurvey/Fourth%20Annual%20Green%20Building%20 Survey %20v3.pdf (March, 2010)

[10]. www.calrecyle.ca.gov California EPA, "Waste Disposal and Diversion Findings for Selected Industry Groups" (2006) Retrieved from

[11]. www.calrecycle.ca.gov/Publications/LocalAsst/Extracts/34106007/ExecSummary.pdf (March,2010)

[12]. Castro-Lacouture, D. Ospina-Alvarado, A. Roper, K. (2008) "AEC+P+F Integration with green Project

[13]. www.cec.org Commission for Environmental Cooperation "Green building in North America" Retrieved from

[14]. http://www.cec.org/Storage/61/5386_GB_Report_EN.pdf (March, 2010)

United States Environmental Protection Agency, Basic Information/Green Building,

[15]. www.epa.gov/greenbuilding/pubs/about/htm, (accessed 4-20-08)

[16]. Wikipedia, Carbon footprint, www.en.wikipedia.org/wiki/Carbon_footprint.com, (accessed 8-09-08)

[17]. Crosbie, Michael J. Ph. D, R.A., "Commercial High-Performance Buildings," Architecture Week, August 30, 2000 Page

[18]. B1.1, www.architectureweek.com/2000/0830/building_1.1.html (accessed 4-6-08)

[19]. Social Butterfly, www.fly4change.wordpress.com/2008/01/10/defining-green-including-the-ftc/, (accessed 4-10-08)

[20]. greenexhibits.org,DefiningGreen,www.greenexhibits.org/begin/defining_green.shtml, (accessed 4-10-08)

[21]. Wikipedia, Green Building, www.en.wikipedia.org/wiki/Green_building.com, (accessed 3-24-08)

[23]. USGBC, LEED for New Construction & Major Renovations, U. S. Green Building Council, Version 2.2, Sustainable Sites

[24]. Materials Characterization Volume 60, Issue 7, July 2009, Pages 716–728

[25]. Design of normal concrete mixes Second edition, D C Teychenné, BSc(Eng), MIStructE R E Franklin, BSc(Eng), MICE, MIHE, H C Erntroy, MSc, FICE, MIStructE

[26]. Preparation of C200 green reactive powder concrete and its static–dynamic behaviors Zhang Yunsheng *, Sun Wei, Liu Sifeng, Jiao Chujie, Lai Jianzhong

[28]. an Overview of the Benefits and Risk Factors of Going Green in Existing Buildings Alev Durmus-Pedini1 and Baabak Ashuri2

[29]. a facility manager's approach to sustainability Received (in revised form): 24th January, 2005 Christopher P. Hodges

[30]. Sustainable Concretes for Insulated Concrete Form (ICF) Construction by Lindsay Moore

[31]. Element Analysis of the Green Building Process Richard E. Zigenfus November 12, 2008

[32]. concrete.union.edu

[32]. www.ce.memphis.edu/1112/notes/project_2/.../ACI_mix_design.pdf

[33]. www.virginiadot.org/VDOT/Business/asset_upload_file13_3529.pdf

[34]. Design and Control of Concrete Mixtures FOURTEENTH EDITION by Steven H. Kosmatka, Beatrix Kerkhoff, and William C. Panarese

[35]. Aggregates from Natural and Recycled Sources Economic Assessments for Construction Applications—A. Materials Flow Analysis *By* David R. Wilburn and Thomas G. Goonan

*****

Sustainable Development of Manufacturing

Nima Bahiraei

UPM University, Serdang, Malaysia

Academic Editor: *Amin Teyfouri*

ABSTRACT: Realization of sustainable development and sustainable production and meeting the related requirements cause massive challenges for the manufacturing industry. The motivation for this study was the assumption that better understanding of the different aspects of sustainable development helps the companies to adapt more sustainable practices. The paper presents a literature review on sustainable development and production with practices related to the respective topics and then summarizes a study conducted within Finnish manufacturing industry. The results in this study are presented in a framework consisting of six categories. For each category the challenges, means and motivation for realization and objectives are presented. The obtained results provide further and in depth information of sustainable development and sustainable production within the Finnish manufacturing industry for both the industry and academy.

Keywords: sustainable development; sustainable production; supply chain

1.0   INTRODUCTION

This monograph deals with environmentally friendly machining. Environmental concerns are gaining importance in every field of engineering. After the ISO 9000 Quality Management System Standards, the ISO 14000 Environmental Management System Standards and the OSHAS 18001 Occupational Safety and Health Assessment Series were published. The ISO 14000 is a set of standards concerning a way an organization's activities affect the environment throughout the life of its products. These activities range from production to ultimate disposal of the product. It includes effects on the environment such as pollution, waste generation, noise, depletion of natural resources, and energy use. The ISO 14000 standards are designed to cover environmental management systems, environmental auditing, environmental performance evaluation, environmental labeling, and life-cycle assessment. The ISO 14000 specifies that an environmental policy, fully supported by senior management, must exist. It outlines the policies of the company, not only toward the staff but also toward the public. OSHAS 18001 is a document for health safety and health management systems. It is intended to help an organization to control occupational health and safety risks. It was developed in response to widespread demand for a recognized standard against which health and safety measures can be certified and assessed. At many places, the word "green" is considered as a synonym of "environmentally friendly." This may be due to the fact that most of the plants are green and convert CO_2 to O_2 needed for the survival of human being. Green color symbols are often used to represent plants and vegetables. This book could also have been named as Green Machining, but this term is in use in a different sense. The machining of ceramic in the unbaked (pre-sintered) sate is called green machining. However, green engineering and green manufacturing are words in vogue to indicate the environmental concerns in engineering and manufacturing, respectively. Manufacturing includes all steps necessary to convert raw materials, components, or parts into finished goods that meet a customer's expectations or specifications. Machining is one type of manufacturing. This chapter gives a brief introduction of environmentally friendly manufacturing in general, and environmentally friendly machining in particular. All other chapters will focus only on environmentally friendly machining.

2.0 SUSTAINABLE DEVELOPMENTS AND SUSTAINABLE MANUFACTURING

2.1 Definitions and Aspects

The most widely accepted definition for sustainable development is to meet the needs of the present consumption without compromising the ability for the future generation to meet their needs was formed by the Brundtland commission. In addition to this, The Lowell Centre for Sustainable Production defines sustainability in manufacturing and production practices as 'the creation of goods and services using processes and systems that are non-polluting, conserving of energy and natural resources, economically viable, safe and healthful for employees, communities, consumers and socially and creatively rewarding for all working people.' Furthermore, as the ability to achieve sustainable development is dependent on both the customer and the producer, both aspects have to be taken into consideration. Hence, sustainable consumption can be defined as 'the use of goods and services that respond to basic needs and bring a better quality of life, while minimizing the use of natural resources, toxic materials and emissions of waste and pollutants over the life cycle, so as not jeopardize the needs of future generations.

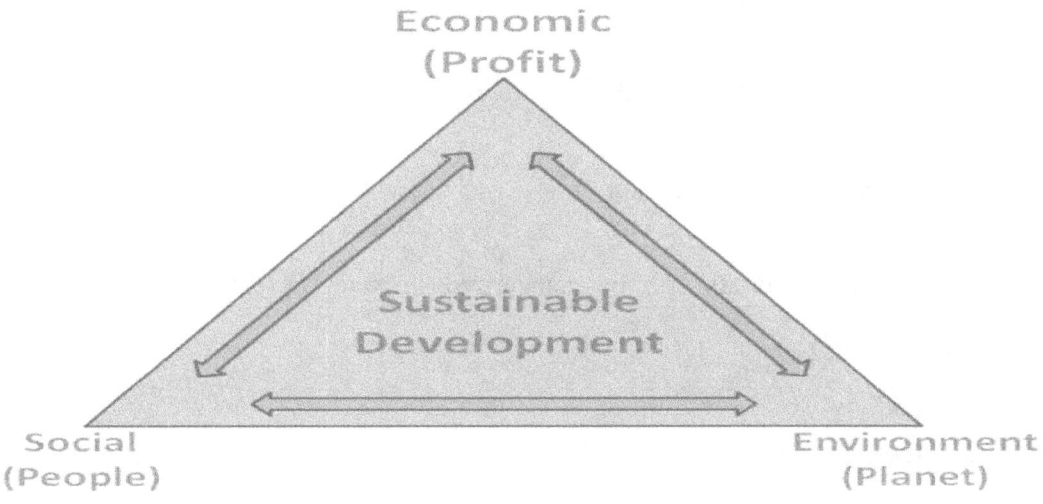

Figure 1: Triple bottom line

Sustainable development is commonly linked to cover three aspects: environmental, social and economic sustainability. These aspects presented in Figure are generally referred as the 'triple bottom line' or 'the three P's', planet, people and profit. Environmental sustainability 'seeks to improve human welfare by conserving the sources of raw material used for human needs and ensuring that the sinks for human wastes are not exceeded, in order to prevent harm to humans'. Manufacturing companies typically seek to reduce the use of materials and energy, and waste and pollution generated. Furthermore, social sustainability can be achieved by assuring people a have reasonable enough share of wealth, influence and safety [9]. From the perspective of the manufacturing companies, this aspect covers issues such as personnel welfare, development and education, occupational health and safety, and compensation. The focus on economic sustainability is to secure both short and long term profitability and economic viability of the company. On the national level the issues related to the economic aspect of sustainable development are for example related to gross domestic product and its development as well as inflation that also affect companies and their operations. The vital point in realizing sustainable development is to balance and develop these three aspects, environmental, societal and economic in unison.

3.0 COMPANIES VIEWS ON SUSTAINABLE

3.1 Development and Sustainable Production

Discussion in sustainable manufacturing can be categorized in many categorized, the categories are as follows:

• Product and Product Design

• Supply Chain

• Production

• Business

• Society

The challenges, means and motivations for realization and objectives of sustainable development and sustainable production in the Finnish manufacturing industry are presented for each of the categories.

3.2 Products and Product Design

Viewing sustainable development from the product and product design aspect, several challenges relate to product send of life management and life cycle assessment. Currently product design is not focused on the sustainability aspect of the product as much as the cost aspect. Products are often viewed as disposable and the current product recovery options are viewed as limited. Furthermore, designing sustainable products is regarded as more time consuming and often the qualities of 'green materials' are not comparable with the traditionally used materials. Also, issues such as reduction of noise and small particle emissions are hard to eliminate from the product without compromising the cost effectiveness. Solutions for overcoming these challenges are seen as developing a systematic product service and recovery strategy. Virtual design tools are seen as an important tool to realize sustainability in the products, in virtual environment many options can be measured, but further development of the design tools needs to take place before it will become feasible solution for product design. Many companies are also evaluating alternative fuels to power their products and their effects on the emission and sustainability of their product. Companies in the Finnish manufacturing industry view product design as an important phase to have an effect on the products sustainability. Especially in products with long-life cycles, the issues in the life cycle should be addressed in an early phase of the product design. The objectives in sustainable product design in the Finnish manufacturing industry are seen as making products modular and updatable. Sustainability in product design is viewed as better life cycle management of the product. Longer term objective for product design is to have an 'eco-product' for which a secondary life cycle can already be planned before the initial life cycle has begun.

3.3 Supply Chain

In many ways several challenges reside in the planning and operating a successful and sustainable supply chain in the Finnish manufacturing industry sector. Partners are often chosen based solely on the cost while other aspects such as environmental impact in the decision play a minor role. Often when a supplier is chosen the goods are delivered in relatively small batches over a long distance. Furthermore, modal transportation is not taking place effectively in the Finnish manufacturing industry. Another issue with the suppliers is that, especially when a supplier is chosen from another country, for example China, the origins of the part are not always made clear for the Finnish company. This raises several ethical questions by the Finnish manufacturing industry, such as worker safety and rights for collective bargaining, work conditions and child labor. The small and medium enterprises consider assessing sustainability in their supply chain operations to be expensive and the time for return of investment for the sustainable practices is considered to be too long to be profitable. Additionally, many of the companies are not sure how they need to address sustainability in their practices. Thus, another issue the companies see is the metrics regarding sustainability in the supply chain, often it is hard for companies to measure their supply chain from a sustainable aspect. The motivation for enhancing sustainability in the supply chain was expressed by the Finnish manufacturing company as a threat that some work might be outsourced to another country than Finland. The means to enhance sustainability in the supply chain are viewed as further collaboration between the companies. For the small and medium enterprises collaboration is considered as extremely important to address the sustainability regulations and issues in the future. Several companies also see committing to sustainability not only in own actions but also in the whole supply chain. The companies selling the goods to the end customers see themselves as the operator to address sustainability and ensure that sustainable practices are being used. However, the pressure to be cost effective is significantly hindering decisions in the corporate strategy to choose the sustainable choice. Several companies expressed willingness to be able to buy the sustainable services from a third party. In the supply chain both long and short-term objectives were presented by the Finnish manufacturing industry. A sustainable supply chain [2]is viewed as short distance transports between the operators. Companies see that it is important for the operators in the supply chain to be located close to each other. Long-term objective for supply chain is viewed as an eco-friendly industrial area where several operators of the supply chain are located and a third party would provide them with services regarding sustainability. Tampere University of Technology's department of production engineering has an ongoing project regarding this called 'CSM-Hotel'. The CSM-Hotel offers operators equipment and facilities on a pay-per-use basis.

3.4 Production

In the production operations the biggest challenges lies in the finances and measurements. The Finnish manufacturing companies experience the costs of a sustainable investment to be far too expensive compared to the benefit gained. Sustainability related investments are seen to be expensive bulk investments; therefore several companies are reluctant to carry out such large-scale projects. Furthermore, the lack of knowledge and understanding in what to do and when to do it increases the reluctance to realize sustainable investments in the production and manufacturing processes and operations. In addition the lack of easily adaptable measurements for small and medium enterprises decreases the support for investing in sustainability. The companies in the Finnish manufacturing industry do not feel a standardized enough set of metrics and indices has been established to support smaller enterprises in realizing sustainability. However, Finnish companies in the manufacturing industry see several benefits in investing to sustainability despite the lack of financial support. With these investments the companies can continue the production in the original location and achieve savings in the production costs. Hence, including sustainability within the bigger future projects is a more feasible way to realize sustainable development in the manufacturing and production processes and operation rather than initiating multiple smaller projects. The paradigm for sustainability in the production and manufacturing field by the Finnish manufacturing industry is to enhance the existing operations to be more efficient and adapting energy conserving equipment and processes in the factory. Several companies see energy recovery from the processes and the overall reduction in the energy use as a way to address sustainability. Increasing the reusability of the used materials and increasing the recyclability of the waste materials generated from the production and manufacturing processes is also seen as a mean to increase sustainability. In addition, the use of packaging materials was mentioned by a number of companies. Reducing the materials used for packaging directly reduces the capital spent on both the packages and transportation, smaller packages can be loaded more efficiently in the transportation and generate less waste. With digital manufacturing several advantages can be achieved in the production and manufacturing processes. With a virtual testing environment the sustainability factor of the planned changes in the production system can be thoroughly tested. The companies in the Finnish manufacturing industry supports automation as a solution to address sustainability, investing in virtual testing of the processes and operations increases the benefit gained from automation. However, realizing the presented things requires commitment from the companies. Currently the companies in the Finnish manufacturing industry aim for short term profits while the focus on the long term plans in the company's strategy and vision does not support realizing the investments. Hence, the companies have to commit themselves in the field of sustainability to realize sustainable production. Finnish manufacturing industry sees that a sustainable production system can be described as effective use of available resources. As a short-term goal the Finnish manufacturing industry aims to develop new production systems and enhance the existing ones to be more sustainable. Over a longer period of time several Finnish companies aim for an eco-factory concept. However, eco factory in Finland is considered to be out of reach especially for the small and medium sized enterprise industry.

3.5 Business

The challenges related to the business activity for the companies in sustainable development and sustainable production is the lack of information and demand from both the customers and the management. The company's strategy and vision is formed to meet the customer's requirements and demands. For most companies the external or internal customer does not demand sustainability related practices or information to the extent it would become business driver. The customers typically choose the product or equipment they buy based on other attributes rather than sustainability. Companies currently are also reluctant to make decision supporting sustainability as there is no demand for it, focusing the resources on the currently important demands from the customer are regarded as more feasible option. Furthermore, the lack of pioneer companies engraving sustainability deeper in the strategy and vision does not encourage the companies in the Finnish manufacturing industry to promote sustainability. Small and medium enterprises are reluctant to become pioneers for sustainable development and sustainable production as the effects of promoting sustainability without concrete demand for it from the customers can lead to loss of sales. Motivation for realizing sustainable development and sustainable production should originate from the company's strategy and vision. However, the current metrics does not give the companies support to lead the change towards sustainable production and sustainability in the company's operations. Further development of standardized and easily adaptable metrics is required. Shifting the company's management model to support sustainability the metrics and indices should be standardized to allow evaluation of progress and benchmarking with other companies as well as setting goals. Majority of the customers do not value the sustainability attributes of the products as highly as the Finnish manufacturing companies wants them to. A number of companies experiences the Finnish brand to give an advantage in selling products which are supporting sustainable development. Furthermore, being the pioneer for sustainability in each industry sector can work as a marketing advantage. Focusing resources on tackling the upcoming challenges is experienced to become an advantage when planning for a longer period of time. The Finnish manufacturing aims to realize business supporting sustainable development and sustainable production. The companies' objectives are to differentiate from the competitors by offering the customers a sustainable choice for the product. Engraving sustainable development and sustainable operations deep in the company's strategy and vision, the companies in the

Finnish manufacturing industry expects to give the public a positive image of the company and its values [3].

3.6 Society

Society places a lot of pressure for the companies by instituting new and tighter regulations. Companies experience politics as both a driver for sustainable development as well as a barrier for business. The pressure to stay competitive with the increasing regulations is considered as one of the bigger challenges for realizing voluntary sustainability in the companies' actions. In addition, the costs of manufacturing a product in Finland compared to manufacturing in cheap labor countries are becoming more expensive. Furthermore, realizing similar practices in Finland compared to a country with less strict regulations for emissions and waste management is not considered to be possible. The companies that have production in cheap labor countries stated that there is no support in realizing similar regulations. Overcoming these challenges is considered to be hard from the companies' perspective alone, hence companies state that support from both the European Union and the Finnish government is required. The current projects to support sustainable development are considered to be taking too much time before they are realized thus the pace for the change in the industry sector should be faster for them to be feasible for the companies. Furthermore, several projects, especially the projects in Finland, do not have a production-oriented perspective and the focus is on the general level issues. Environment is currently experienced in too great extent, thus financial and social sustainability does not have the desired emphasis in the projects. The companies in the Finnish manufacturing industry see realizing sustainability as a joint operation with the governing bodies. The pace for the changes should be set in co-operation with the companies to ensure that it does not jeopardize the business by applying stricter regulations too fast. Certificates are currently experienced as a good way for a company to promote their actions for sustainable development. However, in the manufacturing industry the benefit of the certificate is not seen as important as in some other industry sectors. Finnish manufacturing industry experiences the objects for the social aspect of sustainability as being able to generate jobs throughout the supply chain and supporting local production and manufacturing. However, co-operation with the governments and the industry is required in realizing this. A longer-term objective for the Finnish manufacturing industry is to have a worldwide consensus of the regulations. In future it should not matter where the goods are produced and jobs are generated in places where the workers wish to live.

4.0 CONCLUSION

This paper discussed the current state and future expectations of the Finnish manufacturing industry for sustainable development and sustainable production. Based literature, definitions and aspects as well as means for and advantages of sustainable development were presented. The aim was to clarify the challenges, needs and objectives for realizing sustainability within the industry.

The results are presented in six categories, product and product design, supply chain, production, personnel, business and society. In product design the biggest challenges focus in the products life cycle assessment [4], thus further effort is required in this field. Virtual product design environment is suggested as a mean to realize sustainability in the product. The objective to address sustainability in product and product design is experienced as better life cycle management in the near future. Challenges in the supply chain to address sustainability can be considered as mostly financial. Currently being cost-efficient is more important than being sustainable. Therefore, the companies suggest further collaboration between the operators in the supply chain to achieve sustainability. The Finnish manufacturing industry aims to have short distance 308 M. Tapaninaho et al [1].

transportation in the future and working within close distance between the operators in the supply chain. Finances play a big part in the production operations and processes as well. The companies feel that the big bulk investments to address sustainability are often too expensive to realize, hence grouping the sustainability related issues in the future projects is considered as a more feasible solution rather than carrying out smaller projects. In the production operations and processes the industry aims for both energy and resource efficient production. In Finland the personnel aspect has been in high focus in the recent years with the challenge of aging workforce. The focus within the industry is to address the employee health and wellbeing to further lengthen their careers. However, the importance of other personnel related issues such as retaining the knowhow of the older workers within the company and changing the mindset of the employees to adapt sustainable practices. The companies in the Finnish manufacturing industry wishes to keep the talented individuals within the company and view personal talent as valuable asset for the company. Sustainability is not a big issue in the business-oriented practices such as marketing for the company. The customers demand for sustainability is regarded to be a minor factor when making new sales. However, in future the demand for sustainability is seen to become an important part of the business. Companies in the Finnish manufacturing industry aim to strengthen their brand image by working in the forefront of sustainability. The societal aspect of sustainable development has been in focus along the personnel aspect. Currently the strict legislation places too much pressure especially for the small and medium enterprises to follow. Companies suggest that collaboration with the European Union and the Finnish government is required to set the right pace for realizing the sustainable practices. The industry's objective is to be able to provide jobs for people wishing to live outside the bigger cities.

REFERENCES

- 1.Current State and Future Expectations of Sustainable Development and Sustainable Production in the Finnish Manufacturing Industry.M.Tapaninaho, M. Koho, and S. Torvinen, DOI: 10.1007/978-3-642-27290-5_48, © Springer-Verlag Berlin Heidelberg 2012

- 2.SweeSiongKuik, SevVerlNagalingam, YousefAmer, (2011),"Sustainable supply chain for collaborative manufacturing", Journal of Manufacturing Technology Management, Vol. 22 Iss: 8 pp. 984 – 1001

- 3.Sustainable Manufacturing for Global Value Creation G. Seliger Department of Machine Tools and Factory Management, Berlin University of Technology, Germany, DOI: 10.1007/978-3-642-27290-5_1, © Springer-Verlag Berlin Heidelberg 2012

- 4.Development of an environmental performance assessment method for manufacturing process plans, Zhigang Jiang &Hua Zhang & John W. Sutherland,Int J AdvManufTechnol (2012) 58:783–790, DOI 10.1007/s00170-011-3410-7

*****

Overview of Medical Waste Management in Malaysia

Maryam Khadem Ghasemi[a], Rosnah bt. Mohd. Yusuff[a], Mohd Khairol Anuar bin Mohd Ariffin[a], B.T. Hang Tuah Bin Baharudin[a]

[a] *Department Of Mechanical And Manufacturing Engineering, Universiti Putra Malaysia, 43400 Serdang, Selangor*

Email: *maryamkhadem2009@yahoo.com, rosnahmy@upm.edu.my, khairol@upm.edu.my, hangtuah@upm.edu.my*

Academic Editor: *Shahryar Sorooshian*

ABSTRACT: The risks related with healthcare waste and its management has discovered attention across the world in various events, local and international issues. However, safe management of healthcare waste is necessary, to avoid environmental and public health problems. Appropriate hospital waste management practices should include segregation, collection, storage, transportation, treatment and final disposal. The main purpose of the treatment technology is to clean up waste by destroying pathogens. So, selection of the appropriate treatment method in health-care waste (HCW) management has become a challenge task for the hospital managements especially in developing countries. Various technologies are available for the treatment of healthcare waste namely incineration and non incineration. Expected benefit of the research is to improve the hospital waste management and treatment system from incineration technology to non incarnation by setting of new criteria of the hospital waste

management and comparing between two kinds of technologies and, moreover, to determine how to make the treatment process runs more cost efficiently.

Keywords: Sustainable; healthcare waste management, Multi-criteria decision-making; treatment and disposal.

INTRODUCTION

Since Healthcare centers and hospitals are institutions providing various healthcare services to the community that are places for treating patients, they can also be places to spread disease [1]. Waste is a result of inadequate thinking. Between 75% and 90% of hospital waste is non-risk or "general" healthcare waste, comparable to municipal solid waste (MSW). The remaining 10–25% of hospital waste is regarded as infectious and hazardous, and may pose a variety of health risks [2]. Also, Hospital waste states serious threats to environmental health and requires specific treatment and management prior to its final disposal" [3]. Healthcare and medical waste is a particular waste, which is highly hazardous because of its infectious and toxic characteristics. Furthermore, in healthcare units the direct revelation of waste management workers and members of the public to this type of waste increases the hazard that emerges from their treatment [4]. In addition, the growth of the medical sector around the world over the last decade [5, 6] combined with an increase in utilize of disposable healthcare products has contributed to the large amount of healthcare waste being generated [7, 8]. The waste produced in healthcare can be divided in four main classes:

(1) Hazardous and infectious waste that might contain pathogens, like bloody gloves or used hypodermic needles (HMW-I);

(2) Hazardous waste that can cause injury without infection, such as needle pokes (HMW);

(3) Non-hazardous waste such as non-dangerous chemicals and drugs (MW); and

(4) General solid waste (GSW) comparable to domestic waste [9].

Moreover, the waste created in health care actions "carries a higher potential for infection and injury than any other type of waste. Therefore, wherever it is generated, safe and reliable methods for its handling are essential Inadequate and inappropriate management of healthcare waste may have serious public health consequences and a significant impact on the environment" [10]. The management of hospital waste is of the highest importance due to its potential environmental hazards and public health risks. Comparable to any waste handling, appropriate hospital waste management practices should include segregation, containment, internal collection, storage, transportation (external collection), treatment (particularly for healthcare risk waste), and final disposal [2].

BACKGROUND OF STUDY

Safe management of healthcare waste is necessary, to avoid environmental and public health problems, especially related to transmission of infectious diseases, such as HIV infection and hepatitis. In this respect, healthcare waste producers should develop waste management plans to minimize the risks and overall management cost [11].

Safe disposal of health care waste (HCW) consists of four key stages such as segregation, collection and storage, treatment, transport and safe disposal [12], where national legislation must be followed. Four major categories of HCW recommended for organizing segregation and separate storage, collection and disposal are: sharps, whether infectious or not; non-sharps infectious waste; general waste; and hazardous waste. Collection, storage and treatment of these wastes differ from each other. Incineration, disinfection, sterilization, plasma arc and land filling have been adopted for the treatment of HCW in different parts of the world [13].

Various technologies are available for the treatment of healthcare waste. An understanding of the waste category and its volume is vital before deciding the technology to be adopted. Different waste categories have to be handled differently. HCW treatment technologies, especially for the infectious waste are often classified into burn and non-burn technologies, and have their inherent qualities, demerits and application criteria [13]. There are different treatment processes for clinical wastes namely incineration, autoclaving, Chemical disinfection, Plasma Pyrolysis, and microwaving. According to the treatment studies of medical wastes, about 59–60% of regulated medical wastes (RMWs) are treated through incineration, 37–20% by steam sterilization, and 4–5% by other treatment methods [14, 15, 16]. Non-Incineration Technology includes four basic processes: thermal, chemical, irradiative, and biological. The majority of non-incineration technologies employ the thermal and chemical processes. The main purpose of the treatment technology is to clean up waste by destroying pathogens. Facilities should make certain that the technology could meet state criteria for disinfection [17].

There have been many studies about healthcare waste management. An integrated model which utilized the analytic hierarchy process (AHP) with other systems approaches to establish primary HCWM systems to minimize infection risk of patients and workers within the system developed in developing countries [18]. Some researchers described some of the most common treatment and disposal methods used in the management of infectious health-care wastes in developing countries [19]. Hospital waste management models based on system dynamic to minimize risk to public health have developed [2]. AHP used to a multi-criteria assessment of scenarios on thermal processing of infectious hospital wastes [20]. Healthcare waste treatment/disposal alternatives evaluated by two different multi-criteria decision-making (MCDM) namely ANP and ELECTRE [21]. A fuzzy multi-criteria decision making approach used to assessment of health-care waste treatment alternatives [22]. A new

MCDM technique presented based on fuzzy set theory and VIKOR method to evaluate HCW treatment methods [23].

Hence, according to literature mostly health-care institutions generating the wastes are surveyed through the prepared questionnaires, field research and personnel interviews that are useful in analyzing the current situation in developing countries. On the other hand, there are only a few analytical studies about health-care waste management (HCWM that use decision making tools to implement a comprehensive health-care waste management strategy [22]. Also, because of abovementioned studies their methods have not been able to realize casual relationships and to find the cause and effect group between different networks. Therefore, they concluded the steam sterilization (Autoclave) is the best alternative treatment among others as regards it is not suitable treatment technology for all kind of medical wastes. In addition, for new technologies there is need to obtain the cost of development and licensing and cost of maintenance beside of capital cost and operating cost. Thus, a sensitive study is required to assess more factors and criteria and use interrelationship method to find effective alternative treatment for hospital wastes.

STATEMENT OF PROBLEM

The noninfectious waste and infectious waste are two types of medical waste that each type has completely different of management process. All steps in the medical waste management system are measured an important part of hospital waste management operation. Segregation, handling, transportation, treatment and disposal of infectious waste are the most important steps. Medical waste needs special care in its treatment and disposal because of its hazardous and divers characteristics. Also it largely includes plastic compounds and infectious viruses and microorganisms. It is possible to dispose the noninfectious waste with normal and municipal waste, but infectious waste has to be disposed of in a specific method. The burning of solid and regulated medical waste generated by health care creates many problems. It has been widely known that the incineration of medical waste is one of the major sources of dioxins and furans pollution partly due to the presence of polyvinyl chloride (PVC) products [24, 25].

Nowadays, many developed countries have proper management of hazardous waste with legal requirements. Especially the United States of America (USA) and European countries attend the appropriate operation of hazardous waste management. However, in most developing countries the most commonly treatment technology for HCW is incineration where the waste undergoes combustion under controlled conditions. However, in the recent years the shortcomings of incineration have been largely realized and are no more agreed as environmentally benign. In fact, environmentalists consider incinerators to only change the form of the waste, while retaining the hazards. Essentially, an incinerator burns the waste and leaves behind toxic ash and noxious gases with harmful air pollutants. These emissions are claimed to have serious consequences on worker safety, public health and the environment [13, 26]. So, selection of the appropriate treatment method in health-care waste (HCW) management has become a challenge task for the hospital managements especially in developing countries [19]. In addition, to improve HCW management, some studies have indicated the importance of the use of appropriate techniques for disposal [14, 19, 27]. Also the management and treatment of HCW are gaining more attention with the increasing awareness [22].

This includes the management of hospital waste in Malaysia that according to the notification on scheduled waste by the Department of Environment Malaysia, in 2010 is 42,029.33 million tons that improper management of clinical waste management can menace to the health, safety and environment [28] (Figure 1).

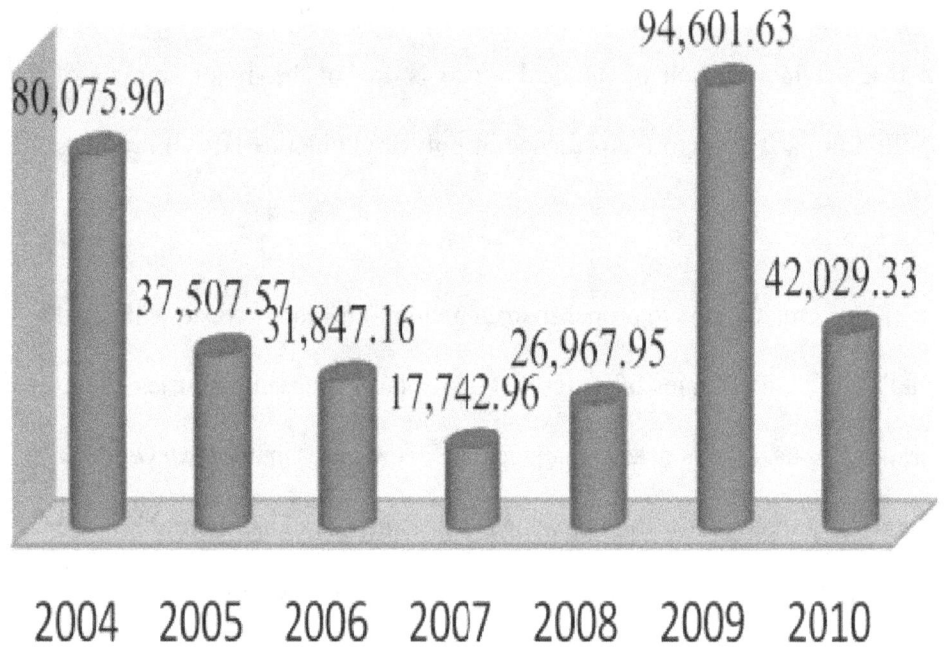

Figure 1: Clinical waste generation in Malaysia from 2004 until 2010 "Metric tons/year" [28]

Furthermore, the estimate average medical waste generation in Malaysia is 1.9 Kg/bed/day that is high compare to other Asian countries except Hanoi [29], It can be an important source of many diseases if not managed appropriately while incineration is mainly used to treat healthcare waste [13] (Figure 2).

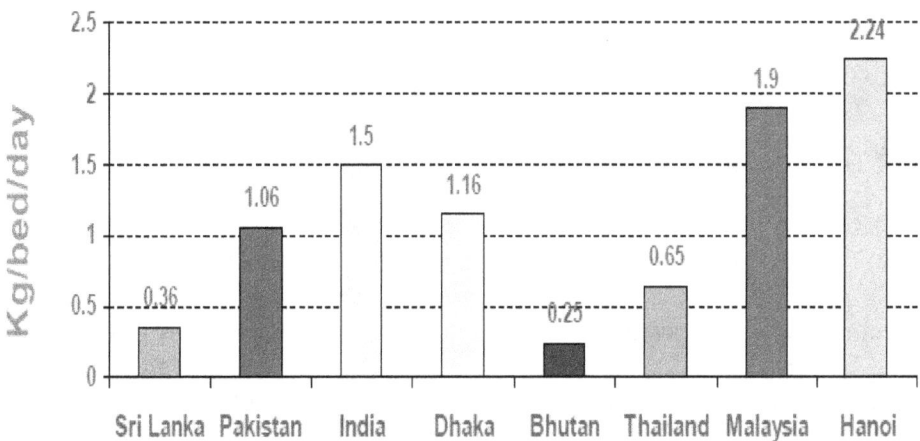

Figure 2: Estimated average healthcare waste generation in some Asian countries [29]

The Ministry of Health confirmed that one of the most common issues faced by clinical waste management is the improper waste segregation at source [30], and Department of Environment further stated that mixture of clinical waste and general waste with together stream is common in hospitals where this problem is an added cost to the concession company, the owner of health care facility as well as the government [31].

Unfortunately, there are not more studies that compare hospital waste treatment between incineration and non incineration. Moreover, there are only a few analytical studies about health-care waste management (HCWM). For this reasons, this study will research and evaluate in hospital waste management systems specially treatment HCW between incineration and non incineration that there is not any suggestion to use single technology to solve the problem of HCW treatment [13] and find out the best non incineration treatment technology due to the type of medical waste to emit fewer pollutants, cost-effective, compact, and reliable and avoid secondary pollution since generate solid residues that are not hazardous.

So, the selection of the effective alternative treatment for hospital waste is a multiple criteria problem such as economical, technical, environmental and social with their sub criteria.

SIGNIFICANT OF STUDY

The first expected benefit is to explore the level of implementation of healthcare waste management practices among Malaysian healthcare industry to use as a principle for resolving problems with the hospital waste management process in Malaysia and to give the recommendations to the Malaysian government healthcare, private healthcare industries, and those interested in improving their existing hospital waste management strategies and standards.

The second expected benefit is to analyze the interactive relation among various elements in healthcare waste management system to find whether the elements impact to each other to develop the best steam of healthcare waste management with non incineration treatment.

The third expected benefit is to use to evaluate of healthcare waste treatment alternative and criteria by applying Multi-criteria decision-making methods (MCDM). In general the multi-criteria decision method, as well as the conclusions and remarks of this study can be used as a basis for future planning and anticipation of the needs for investments in the area of medical waste management.

REFERENCES

[1] Borg, M.A, 2007. Clinical waste disposal-getting the fact right. Journal of Hospital Infection, 65(2): 178-180.

[2] Chaerul, M., M. Tanaka and A.V. Shekdar, 2008. A system dynamics approach for hospital waste management. Waste Management, 28(2): 442-449.

[3] Hassan, M.M., S.A. Ahmed., K.A. Rahman and T.K. Biswas, 2008. Pattern of medical waste management: existing scenario in Dhaka City, Bangladesh. BMC Public Health, 8(1): 36.

[4] Tsakona, M., E. Anagnostopoulou and E. Gidarakos, 2007. Hospital waste management and toxicity evaluation: a case study. Waste Management, 27(7): 912-920.

[5] World Health Organization (WHO), 2002. Basic Steps in the Preparation of Health Care Waste Management Plans for Health Care Establishments (Accessed online at http://www.healthcarewaste.org).

[6] Karamouz, M., B. Zahraie., R. Kerachian., N. Jaafarzadeh and N. Mahjouri, 2007. Developing a master plan for hospital solid waste management: A case study. Waste Management, 27(5): 626-638.

[7] Da Silva, C., A. Hoppe., M. Ravanello and N. Mello, 2005. Medical wastes management in the south of Brazil. Waste Management, 25(6): 600-605.

[8] Patwary, M.A., W.T. O'Hare., G. Street., K. Maudood Elahi., S.S. Hossain and M.H. Sarker, 2009. Quantitative assessment of medical waste generation in the capital city of Bangladesh. Waste Management, 29: 2392-2397.

[9] Giacchetta, G. and B. Marchetti, 2013. Medical waste management: a case study in a small size hospital of central Italy. Strategic Outsourcing: An International Journal, 6(1): 65-84.

[10] Prüss, A, E. Giroult and P. Rushbrook, 1999. Safe management of wastes from health-care activities. Geneva: World Health Organization, p. 1-230.

[11] Graikos A, E. Voudrias, A. Papazachariou, N. Iosifidisb and M. Kalpakidoub, 2010. Composition and production rate of medical waste from a small producer in Greece. Waste Management, 30: 1683–1689.

[12] World Health Organization (WHO), 2008. Healthcare Waste and its Safe Management (Accessed online at http://www.healthcarewaste.org).

[13] Prem Ananth, A., V. Prashanthini and C. Visvanathan, 2010. Healthcare waste management in Asia. Waste Management, 30(1): 154-161.

[14] Lee, B.K., M.J. Ellenbecker and R. Moure-Ersaso, 2004. Alternatives for treatment and disposal cost reduction of regulated medical wastes. Waste Management, 24(2): 143-151.

[15] Park, H.S. and J.W. Jeong, 2001. Recent trends on disposal technologies of medical waste. Journal of Korea Solid Wastes Engineering Society, 18 (1): 18–27.

[16] Hyland, R.G., D.A. Drum and M. Bulley, 1994. Disposal medical wastes. In: 87th Annual Meeting and Exhibition, Air & Waste Manage. Assoc., Paper No. 94-RP-123B.02, Cincinnati, OH, 19–24 June.

[17] Katoch, S.S, 2007. Biomedical waste classification and prevailing management strategies. In proceedings of the International Conference on Sustainable Solid Waste Management, 5 - 7 September 2007, Chennai, India. Pp.169-175.

[18] Brent, A.C., D.E. Rogers.,T.S. Ramabitsa-Siimaneand M.B. Rohwer, 2007. Application of the analytical hierarchy process to establish health care waste management systems that minimise infection risks in developing countries. European Journal of Operational Research, 181(1): 403-424.

[19] Diaz, L., G. Savage and L. Eggerth, 2005. Alternatives for the treatment and disposal of healthcare wastes in developing countries. Waste Management, 25(6): 626-637.

[20] Karagiannidis, A., A. Papageorgiou., G. Perkoulidis., G. Sanida and P. Samaras, 2010. A multi-criteria assessment of scenarios on thermal processing of infectious hospital wastes: A case study for Central Macedonia. Waste Management, 30(2): 251-262.

[21] Özkan, A. 2013. Evaluation of healthcare waste treatment/disposal alternatives by using multi-criteria decision-making techniques. Waste Management and Research, 31(2):141-149.

[22] Dursun, M., E.E. Karsak and M.A. Karadayi, 2011. Assessment of health-care waste treatment alternatives using fuzzy multi-criteria decision making approaches. Resources, Conservation and Recycling, 57: 98-107.

[23] Liu, J., S.F. Liu., P. Liu., X.Z. Zhou and B. Zhao, 2013. A new decision support model in multi-criteria decision making with intuitionistic fuzzy sets based on risk preferences and criteria reduction. Journal of the Operational Research Society, 64 (8): 1205-1220.

[24] Vesilind, P.A., W.A. Worrell., D.R. Reinhart, 2002. Solid waste engineering: Brooks/Cole Pacific Grove.

[25] Jang, Y.C., C. Lee., O.S. Yoon and H. Kim, 2006. Medical waste management in Korea. Journal of Environmental Management, 80(2): 107-115.

[26] Healthcare Without Harm (HCWH), 2001. Non-Incineration Medical Waste Treatment Technologies: A Resource for Hospital Administrators, Facility Managers. Health Care Professionals, Environmental Advocates, and Community Members, Washington, DC, USA (accessed online at www.noharm.org/nonincineration).

[27] Rogers, D.E. and A.C. Brent, 2006. Small-scale medical waste incinerators–experiences and trials in South Africa. Waste Management, 26(11): 1229-1236.

[28] Omar, D., S.N. Nazli and S.A/L. Karuppannan, 2012.Clinical Waste Management in District Hospitals of Tumpat, Batu Pahat and Taiping, Procedia - Social and Behavioral Sciences, 68: 134–145.

[29] Visvanathan, C, 2006. Medical waste management issues in Asia. Paper presented at the International Conf "Asia 3R Conference. 30 October – 1 November, 2006, Tokyo, Japan.

[30] Ministry of Health (MOH), 2009. Health Care Waste Status Report. Putrajaya: Engineering Services Division, Ministry of Health, Malaysia.

[31] Department of Environment (DOE), 2009. Guideline on the handling and management of clinical wastes in Malaysia, 3rd edition, Putrajaya: Government Printers.

*****

From Idea to Enterprise: The Business Concept Statement

Yap Chui Yan[a], Lee Chia Kuang[a], Chong Kwok Feng[b], Liu Yao[a]

(Faculty of Technology [a], Faculty of Industrial Science &Technology [b], Universiti Malaysia Pahang, Tun Razak Highway, 26300 Kuantan, Pahang, Malaysia)

Academic Editor: *Shahryar Sorooshian*

ABSTRACT: This paper provides information on preparing a business concept statement in the initial stage of starting a business enterprise based on a university research case study. The first part of the paper covers the definition and importance of a business concept. The second part explains the critical contents that should be included in an effective business concept statement. Finally, the W2W Technologies case study at the end of this paper shows an example of how a business concept statement might look like.

Keywords: Business Concept, Commercialization, Start-up Venture

1.0 INTRODUCTION

Today, universities are increasingly engaged in commercialization, technology transfer, and lab to market projects. These projects provide researchers with an opportunity for weighing, refining, and starting up a prospective enterprise from their research invention. However, for researchers, it is often easier to come up with a variety of research ideas or inventions and more difficult to actually put forth their lab ideas and inventions in the market. This is usually the case because it takes a lot of time, great commitment, and farsightedness to develop the research idea into an innovative business idea that appeals to an investor. So how does a researcher turn the research idea into a commercial reality? The answer is to begin with a business concept statement. This paper offers information on what is a business concept statement, what are the contents of a sound and convincing business concept statement, and how a business concept statement might be presented.

2.0 DEFINITION AND IMPORTANCE OF A BUSINESS CONCEPT STATEMENT

A proposed business venture undergoes three phases of development: idea, concept, and plan. A business concept is basically a bridge between a business idea and a business plan [1]. For example, a business idea may be a newly invented product or process, an improvement to an existing product or process, or a new form of distribution. As the idea is transformed into a concept statement, the researcher cum entrepreneur tests out the business idea to see if there are customers for it and how big the market might be. Once the business concept statement hurdle is cleared, the entrepreneur will have at least thought through some potential challenges and strategies and learned more about the market and potential profitability and financing of the proposed business. Following this, the more detailed work of business planning and implementation takes place.

Preparing a concept statement is necessary because it helps the entrepreneur to (1) lay out clear specifics regarding the proposed venture, (2) evaluate whether the proposed venture is feasible or not, and (3) unearth critical components of a venture that might need to be more thoroughly addressed in an extensive and detailed business plan. In addition, a clear business concept statement also enables the entrepreneur to succinctly describe the precise prospect of the business to team members, investors, partners, suppliers, and customers [2].

3.0 CRITICAL CONTENTS OF A BUSINESS CONCEPT STATEMENT

The five critical questions addressed in an effective business concept statement can be summarized as follows:

1. What is the problem; or what is the pain of the customer?
2. What is the solution; or how does the proposed business overcome the pain of the customer?
3. What is innovative about the business? How unique is the business? Can it be protected? Can a competitor simply copy it?
4. What is the market size and primary target customer?
5. Can the business make money? What are the costs involved and what price can be asked?

What is the problem; or what is the pain of the customer?

Where does the idea or opportunity for a business come from? A good starting point is usually to look into a current or existing problem. Bear in mind that the problem is not just any problem or pain in the society. The problem or pain should relate directly to a group of customers who are willing to pay for having their problems resolved or pain eliminated [3]. So, the clearer a business is able to understand and describe customers' problem and pain, the greater the business potential and opportunity.

What is the solution; or how does the proposed business overcome the pain of the customer?

Many start-ups fail because customers do not buy the product or service. The reason is likely because customers do not understand the advantage of using the product or service. Plus, it is difficult to entice customers away from what they are used to unless they see a clear advantage for doing so. In describing the business concept, thus, entrepreneurs should overtly state how customers will benefit from using the product or service and show how the benefits are of greater value compared to alternative solutions in the market.

What is innovative about the business? How unique is the business? Can it be protected? Can a competitor simply copy it?

Investors look for innovation and uniqueness in a business. The business concept should include information about the uniqueness of the product or service, or how it is differentiated from other, similar products or services. If possible, describe how the uniqueness of the business can be protected. Ultimately, it is important for investors to understand the competitive advantage of the business and that the business is protected against threats of new entrants [4].

What is the market size and primary target customer?

With regard to the market, investors are particularly interested in two questions: How large is the market and who are the primary customers [5]. Although detailed market analyses and figures are required only at business planning stage, at the business concept stage it is necessary to estimate the size of the market. This can be done by using reasonable assumptions based on well-founded data and sources. In specifying the target customers, the entrepreneur should show how the business offers particular benefits to precisely these customers. In the W2W Technologies example, customers who are most receptive towards buying and using UMPacitor will be those that are keen on supporting recycling and green technologies.

Can the business make money? What are the costs involved and what price can be asked?

Detailed financial analysis is not required at the business concept stage. It is expected, however, for the entrepreneur to at least report a rough estimate of cost and revenue, gross profit margin, and the various streams of revenue, where appropriate. A rule of thumb for high-growth business (i.e., a business that has sales of around RM25 million, employs at least 100 people, and operates nationally) is that it should achieve gross profit margin of 40-50 percent during the first five years of start-up [2].

4.0 EXAMPLE OF A BUSINESS CONCEPT STATEMENT

The following section presents an example of a business concept statement for W2W Technologies, a proposed business enterprise based on an actual university research case study.

W2W Technologies Business Concept Statement

The Problem

Every governmental and business organization that advocates sustainability and green technology consider recycling as one of the basic components in a sustainable programme. Many will want to buy and use recycled products in support of efforts to fight against waste, conserve resources, prevent pollution and protect the environment. Available recycled products for industrial use, however, is either too costly or is still lacking despite a clear need for them.

The Solution

To address the growing need for recycled products for industry use, W2W Technologies will produce and sell UMPacitor, which is an energy storage super capacitor made from recycling spent battery. All disposed batteries are hazardous waste and contribute to the growing amount of toxins leaking from landfills. Customers who buy and use UMPacitor will enjoy three benefits. First, they will help to boost battery recycling programmes throughout the country. Second, the customers will earn a better reputation as an organization that supports "go green" efforts. Third, given that the UMPacitor is comparable to commercially available supercapacitors in the market in terms of functionality and price, the transition to using UMPacitor should be smooth for the customers.

The Innovation

W2W Technologies is innovative in several ways. First, in terms of the materials used in production. W2W Technologies will develop UMPacitor using recycled materials extracted from dismantled spent batteries. Research shows that there are useful materials in spent batteries, namely carbon, manganese, cobalt and lithium. The materials will be treated for activation and then fabricated into UMPacitor. The development and design process of UMPacitor will be filed to acquire legal date stamp and will be used for preliminary patent filings. Another innovative aspect of W2W Technologies is its participation in battery recycling efforts. This will likely lead to unique collaboration among various environment friendly partners. The relationship capital developed through such collaboration will be difficult to imitate by other enterprises.

The Market Size and Target Customers

IDTechEx Research 2013 forecasts for supercapacitors show that the global market for supercapacitors will be worth over US$3 billion (approximately RM9 billion) in 2018 with a robust 30% annual growth [6]. Although potential customers can be found in industries producing automotive subsystems, consumer electronics, renewable energy systems, and portable power tools, the initial primary customers that find UMPacitors particularly attractive will be those that support "go green" efforts as part of their sustainability and corporate social responsibility projects.

Business Model/Revenue Mechanism

W2W Technologies will focus on developing a range of UMPacitors for various industrial applications. It will work with suppliers (e.g., Duracell, Energizer, Panasonic, Eveready) for battery recycling and partners in producing and selling UMPacitors. The revenues will be generated from sales of UMPacitors and licenses to third party manufacturers. Initial financial analyses show that capturing a 0.0025% (approximately RM23 million) of the total market and operating profit margin of 30-50% should be possible for W2W Technologies in the first five years of start-up.

5.0 ACKNOWLEDGMENTS

The authors thank the Malaysian Technical University Centre of Excellence (MTUN COE) for funding RDU 121214.

REFERENCES

1. Stangler, D., 2012, "How to Turn an Idea into a Start-up: Begin with a Business Concept Statement," retrieved from http://www.forbes.com/sites/kauffman/2012/06/11/how-to-turn-an-idea-into-a-start-up-begin-with-a-business-concept-statement/, accessed on 12 May 2013.

2. Kubr, T., Marchesi, H., Ilar, D. and Kienhuis, H., 1998, "Starting Up: Achieving Success with Professional Business Planning", McKinsey & Company, Inc., Amsterdam.

3. Byers, T. H., Dorf, R. C. and Nelson, A. J., 2011, "Technology Ventures: From Idea to Enterprise" (3rd Ed.), McGraw-Hill., New York.

4. Mullins, J., 2010, "The New Business Road Test: What Entrepreneurs and Executives Should Do Before Writing a Business Plan" (3rd Ed.), FT Prentice Hall, Harlow.

5. Scott, S. A., 2005, "Finding Fertile Ground: Identifying Extraordinary Opportunities for New Ventures", Prentice Hall, Upper Saddle River.

6. IDTechEx. 2013, "Change of Leadership of the Global Market Value of Supercapacitors?" retrieved from http://www.idtechex.com/research/articles/change-of-leadership-of-the-global-market-value-of-supercapacitors-00005344.asp, accessed on 31 August 2013.

*****

Business Feasibility Assessment on Energy Storage Capacitor: A case of Commercializing University Research.

Lee Chia Kuang[a], Liu Yao[a], Chong Kwok Feng[b], Yong Ying Mei[c]

(Faculty of Technology[a], Faculty of Industrial Science & Technology[b], Centre for Modern Language and Human Science[c], Universiti Malaysia Pahang, Tun Razak Highway, 26300 Kuantan, Pahang, Malaysia)

Academic Editor: *Shahryar Sorooshian*

ABSTRACT: The main aim of this paper is to discuss a case of commercializing university research. Both importance of conducting a feasibility assessment and the concept of waste to wealth are brought up. In the data collection section, a standard business concept feasibility assessment has been carried out with the principal researcher to identify the strengths and weaknesses in commercializing research on energy storage capacitor. Following this assessment, we proposed several strategies to consolidate the strengths and overcome the weaknesses, and bring the research closer to commercialization and greater heights.

Keywords: Commercialization, Energy Storage Capacitor, Business Concept Feasibility

1.0 INTRODUCTION

Feasibility study is basically an analytical tool used in the industry before the implementation of any project. A general project feasibility study is essential and vital during the appraisal stage. It reveals potential impacts of the proposed business. The preparation of project feasibility study generally is to assess options, and address alternatives to the client on whether a project, or a business should be implemented based on evaluations of feasible criteria. In other words, a project feasibility study is conducted mainly to determine and decide whether a project is profitable and realistically be achieved, before the project even takes place [1]

Business Concept Feasibility however differs from a general project feasibility assessment. It identifies strengths and weaknesses in a business concept, and identifies major business risks. Assessed personnel can determine the level of risk taken in commencing a business, and take constructive steps in the assessed criteria for better improvements.

2.0   WASTES TO WEALTH CONCEPT

Given the fact that disposal of spent batteries damage the environment due to its hazardous metallic content, the principal researcher has researched and recycled spent battery for an energy storage super capacitor. The proposed energy storage super capacitor has tremendous potential in charge storage ability upon appropriate recycling and activation. This concept of from "waste to wealth" would enhance a cleaner environment by reducing he contemporary storage mechanism in energy consumption and minimize pollution.

This state of the art super-capacitor has large operating voltage of 3V, high reversible recharge ability of more than 1000 times, thin in size. The novelty of this product lies on its reversible recharge ability and operating voltage, compared to contemporary batteries that need to be landfilled or incinerated.

The potential market identified so far includes automotive subsystems, consumer electronics, short term UPS and telecommunication industries, and usable for small medium enterprises of renewable energy systems and portable power tools.

3.0 BUSINESS CONCEPT FEASIBILITY ASSESSMENT

To accomplish the targets of this study, the business viability of the energy storage Super capacitor was conducted by adopting the FastTrac® TechVenture™ business feasibility assessment tool. This assessment tool identifies the major criteria that can make or break the success of a business, encompassing the aspects of the product or service feasibility, market feasibility and financial feasibility. And for each criterion, there are several sub-criterions; and for each sub-criterion, there are five descriptions denoted from 1 through 5. By evaluating your business concept against these criteria, you can determine the levels of risks you might be taking in starting that business and also develop benchmarks for improvement in the product/service, the market, and the financial aspects of the business. The major criteria are demonstrated in Figure1 below:

Criteria	Sub-criteria	1	2	3	4	5
PRODUCT OR SERVICE FEASIBILITY	Customers perceive a need for product/service.	☐	☐	☐	☐	☐
	Product/Service is ready to sell.	☐	☐	☐	☐	☐
	Product/Service has unlimited life.	☐	☐	☐	☐	☐
	Product/Service is unique and protectable.	☐	☐	☐	☐	☐
	Product/Service is not regulated by the government.	☐	☐	☐	☐	☐
	Product/Service line has expansion potential.	☐	☐	☐	☐	☐
	Product/Service has no liability risk.	☐	☐	☐	☐	☐
MARKET FEASIBILITY	Market can be recognized and measured.	☐	☐	☐	☐	☐
	Existing competition has identifiable weaknesses.	☐	☐	☐	☐	☐
	Distribution system is established and receptive.	☐	☐	☐	☐	☐
	Customers purchase frequently.	☐	☐	☐	☐	☐
	Business has great news value.	☐	☐	☐	☐	☐
FINANCIAL FEASIBILITY	Funding is easily obtained.	☐	☐	☐	☐	☐
	Revenue stream is continuous.	☐	☐	☐	☐	☐
	Money is collected prior to sales.	☐	☐	☐	☐	☐
	Hiring and retaining employees is easy.	☐	☐	☐	☐	☐
	Inventory/Service providers are dependable.	☐	☐	☐	☐	☐
	Gross margin is 100 percent.	☐	☐	☐	☐	☐
	Legal problems do not exist.	☐	☐	☐	☐	☐
	Wealth is generated through exit strategy.	☐	☐	☐	☐	☐

Figure1: Business Concept Feasibility Assessment [2]

To complete this assessment, the research team was requested as respondent to review the criteria and their corresponding descriptions first; then, choose the description that most closely matches their situation and mark the description's number (1 through 5). After rating all the criteria, the score was totaled against the full 100 points ("5" * "20" criteria).

To note, no score guarantees a feasible business. Generally, for a business concept to be feasible it must achieve at least one-half of the possible points overall (≥ 50) and at least one-half of the possible points in any of the three feasibility areas—product or service (≥ 17.5), market (≥ 12.5) and financial (≥ 20). A score of less than 3 for any individual criterion indicates a weakness in the business concept. And base on those identified weaknesses, further action suggestions could be proposed.

4.0 FINDINGS

Depicted in Table 1, the weaknesses that have been identified are nearly nine fold. There are **Four (4)** weaknesses identified under Product /Service Feasibility; **One (1)** weakness under market feasibility; and **Four (4) weaknesses** under Financial Feasibility.

Under ***Product and Service Feasibility Assessment***, the research team felt that the future customers need to be educated about the products and services. The team also felt that there will be several future predicaments that will impede his product sales. There will be limited expansion potential for his current product and he is unsure with the risks in selling his products and services.

Under ***Market Feasibility Assessment***, the team has not yet established any distribution system that is able to carry their products.

Under ***Financial Feasibility Assessment***, the research team is not sure with the amount of money required for starting the business. In addition to that, the team has a poor concept of credit management. Added with lack of manpower, the team has almost no idea where and how to get the inventories and supplies needed.

Overall, the research team scored a relative low mark in the business concept assessment.

Table 1: Business Concept Feasibility Assessment Scoring [2]

PRODUCT OR SERVICE FEASIBILITY	1	2	3	4	5
Customers perceive a need for product/service.		✓			
Product/Service is ready to sell.		✓			
Product/Service has unlimited life.				✓	
Product/Service is unique and protectable.			✓		
Product/Service is not regulated by the government.				✓	
Product/Service line has expansion potential.		✓			
Product/Service has no liability risk.		✓			
MARKET FEASIBILITY	1	2	3	4	5
Market can be recognized and measured.			✓		
Existing competition has identifiable weaknesses.				✓	
Distribution system is established and receptive.	✓				
Customers purchase frequently.			✓		
Business has great news value.			✓		
FINANCIAL FEASIBILITY	1	2	3	4	5
Funding is easily obtained.		✓			

			✓		
Revenue stream is continuous.			✓		
Money is collected prior to sales.	✓				
Hiring and retaining employees is easy.		✓			
Inventory/Service providers are dependable.		✓			
Gross margin is 100 percent.				✓	
Legal problems do not exist.					✓
Wealth is generated through exit strategy.				✓	

5.0 PROPOSED STRATEGIES

To mitigate product risks, we proposed the research team to approach and speak to a patent attorney. Preliminary patent search is essential and the likelihood of such technology to receive a patent need to be assessed at the first place. To foster customer's awareness and sales outcome, we proposed the research team to perform working prototypes with consumers through exhibitions, fairs, and possible industrial booths sections.

To understand the market demands on green products, especially on recycled batteries, we propose to study the intentions of customers to purchase, and the determinants that would affect them effectively. To influence and understand consumer behavior is never an easy task, and we tend to suggest a further investigation to be done by using **Planned Behavior Theory**. This would enhance both the authors and the research team a better understanding of the behavior of the potential users before any distribution system is being established.

6.0 ACKNOWLEDGMENTS

The authors thank the Malaysian Technical University Centre of Excellence (MTUN COE) for funding RDU 121214.

REFERENCES

23. Hyari, K. and Kandil, A, 2009, "Validity of Feasibility Studies for Infrastructure Construction Projects", Jordon Journal of Civil Engineering, 3(1), 66-77.
24. Ewing Marion Kauffman Foundation, 2006, " Business Concept Feasibility Assessment", FastTrac® TechVenture™, page 1-5.

*****

Elements of Total Quality Management

Shahryar Sorooshian[1], Tan Seng Teck[2]

[1]*Faculty of Technology, University Malaysia Pahang, Malaysia*

[2] *School of Business, INTI International College Subang, Malaysia*

Acedemic Editor : *Siti Aissah Mad Ali*

ABSTRACT: This Paper is in format of hypothesis model development. In this paper, based on the extensive literature review, a conceptual model developed; in which three factors of total quality management (TQM) are highlighted. TQM Critical Success Factors (CSF), TQM Quality tools (QT), and TQM Performance Measures (PM) structures the model. The developed model Probe the interaction of the three prongs (CSF, QT, and PM). Also suggestions for further researches are listed to guide scholars to modify our conceptual model.

Keywords: Total quality Management, Critical Success Factors, Quality tools, Performance Measures, Interaction modeling.

1.0 INTRODUCTION

The significance of TQM in current years has been increased both theoretically and practically. In that regard, paying attention to principles such as Tools and Technique, Critical Success Factors, and Performance Measures and their relationship with each other has been noticeably important and investigated. The current concern in total quality management (TQM) implementation is its success.

The researchers' results show the key success or failure of TQM. Different results have been achieved, and their diversities are due to differing assumptions, conditions and research methods. However, generally rewarding results have been obtained.

According to the literature, total quality management (TQM) was driven by three main elements: critical success factors (CSFs), quality tools and techniques (QTs) and performance measures (PMs). Most researches on TQM have examined the relationships between two elements only, such as effect of quality tools (QT) or critical success factors (CSFs) on performance measures, or TQM as a single construct with performance measures (PM) or consider some factors as a mediator between two elements.

The debate is the manner that Performance Measures, Critical Success Factors, and implementation of Quality Tools can be playing their roles as the driving force of TQM activities, both individually and with respect to the other elements involved to bring successive results for companies, which have used TQM as their management philosophy.

When TQM fails, it is not because the basic concept is faulty; observers have argued that the problem lies either with the failure to implement fully all the keys TQM practices or with the absence of complementary assets that must be combined with TQM to achieve a competitive advantage [1-2]. Inappropriate performance measurement can act as a barrier to TQM implementation [3-6]. On the other hand, what has been missing from the available literature is an assessment of how tools of quality have affected TQM[7].

Many articles have been written about the elements of TQM and the approaches taken to ensure a successful implementation of TQM; but, as widely discussed and applied as TQM is, the issues continue to remain complex and somewhat clouded. Few academic studies have attempted to identify the elements that are critical success factors for the development and implementation of a TQM program [7-8].

One of the most significant current discussions in total quality management (TQM) practices is achieving success from implementing of TQM by plenty of articles in this regard. Recent researches on TQM mostly have examined the relationships between the elements of TQM practices, while being a lack of examination on interaction of the three main elements.

This review shows, on the one hand, that there have been numerous studies analyzing the critical factors for successful quality management implementation and its influence upon performance [9-11], and on the other, which techniques and tools might be best suited for quality improvement. In this latter case, there is a major gap in research in this area, because there are few studies which have verified if the use of these techniques and tools improves the TQM level and if it has an influence upon performance.

Thus, considering that: (a) an effective TQM program has positive effects upon operating performance [11], (b) the use of these techniques and tools is vital to support and develop the quality improvement process [7, 12-14].

Experience has shown that some firms fail when they implement TQM [15-16] because the implementation of TQM cannot be successful without the use of suitable quality management methods [17-19] such as tools and techniques for quality. According to this view, the management system of TQM may only have a positive effect on performance if a technical system has also been established [20]. In addition, these techniques, amongst others, are important for business survival and continuation [7, 21].

Fundamentally, the emphasis on measurement is predicated on the belief that measuring results can lead to better results [22] and can increase the chances of a TQM program's success [23].

The practice of measuring is a tangible testament to a firm's true commitment to the tenets of TQM. While many TQM programs "generate more enthusiasm than tangible improvement", this is often because of a failure to link programs with results[24].

2.0 MODEL DEVELOPMENT

The purpose of this study is developing a conceptual model base on three main elements of TQM to be prepared to define the most probably best implementation of TQM practices.Our proposed model which is shown in figure 1, is based on the need to align strategies (CSFs) actions (QTs) and performance measures (PM) identified by Dixon et al [25].

In the following graphically conceptual model, we draw three concepts of TQM, which are CSFs, PMs, QTs; and try to highlight the interaction (direct and indirect relationship) of these three factors on each other and also try to find the TQM alignment base on these three factors. Also the model shows the need to identification, and to list the elements of CFS, QT, and PM.

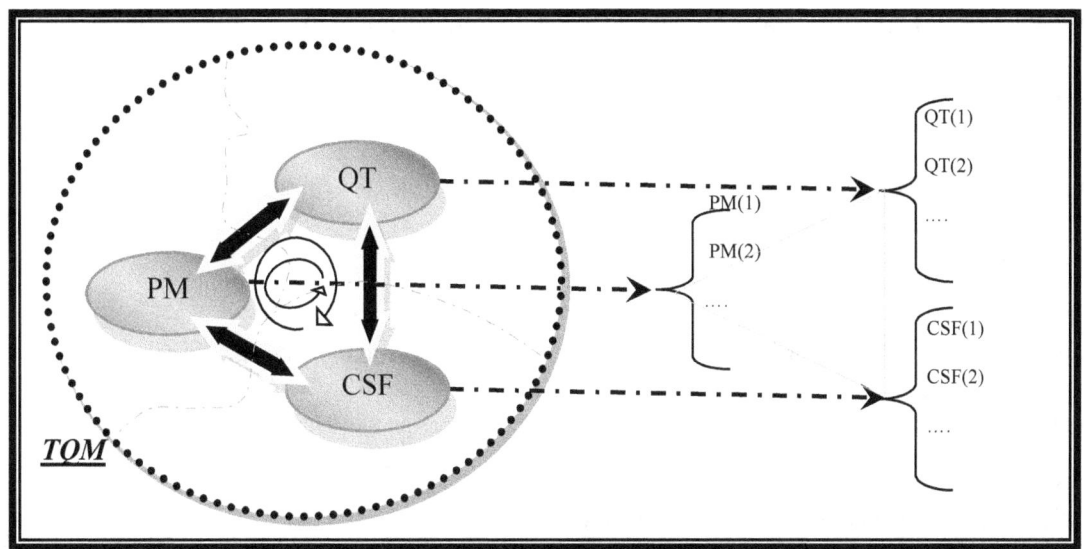

Figure 1: TQM three-pronged model

The developed TQM three-pronged model was tested in an e-survey by sending the model through an email to some of TQM academic experts, and they were asked to judge the model. The list of experts were selected from academic staff of graduate school of management in university Putra Malaysia [26]. Feedback supports our hypnotized conceptual model.

3.0 DISCUSSION

This study highlights the interactions and alignment between TQM three variables and determines the most effective method in TQM implementation. The proposed three-pronged approach model which can define an effective method for industries/organizations to achieve their goals from TQM implementation is expected. It is hoped an implementation of TQM will require less time and cost and increased success potential by using the model.

Following the developed conceptual model may lead organization to optimize the TQM implementation cost, duration and finally effectiveness of TQM implementing. Base on our developed conceptual model; further research suggestion for scholars, are:

1) To measure the role of Quality Tools implementation to improvement of Performance Measures and vice versa.

2) To measure the role of critical success factors identification to improvement of Performance Measures and vice versa.

3) To measure the role of Quality Tools implementation to critical success factors identification and vice versa.

4) To classify Performance Measures, Quality Tools and CSFs.

5) And the most important is to come up with interrelationship modeling for CSFs, Quality Tools, and Performance Measures.

REFERENCES

1. Hackman, J.R. and R. Wageman, Total quality management: Empirical, conceptual, and practical issues. Administrative Science Quarterly, 1995. 40(2): p. 309.

2. Douglas, T.J. and W.Q. Judge, Total Quality Management Implementation and Competitive Advantage: The Role of Structural Control and Exploration. The Academy of Management Journal, 2001. 44(1): p. 158-169.

3. Kanji, G.K., Measurement of business excellence. Total Quality Management, 1998. 9(7): p. 633-643.

4. Ritchie, L. and B.G. Dale, Self-assessment using the business excellence model: A study of practice and process. International Journal of Production Economics, 2000. 66(3): p. 241-254.

5. McAdam, R. and A. Bannister, Business performance measurement and change management within a TQM framework. International Journal of Operations & Production Management, 2001. 21(1): p. 88-107.

6. Chang, H.H., The influence of continuous improvement and performance factors in total quality organization. Total Quality Management and Business Excellence, 2005. 16(3): p. 413-437.

7. Tari, J.J. and V. Sabater, Quality tools and techniques: Are they necessary for quality management? International Journal of Production Economics, 2004. 92(3): p. 267-280.

8. Antony, J., et al., Critical success factors of TQM implementation in Hong Kong industries. The International Journal of Quality & Reliability Management, 2002. 19(5): p. 551.

9. Saraph, J.V., P.G. Benson, and R.G. Schroeder, An instrument for measuring the critical factors of quality management. Decision Sciences, 1989. 20(4): p. 810-829.

10. Powell, T.C., Total quality management as competitive advantage: A review and empirical study. Strategic Management Journal, 1995. 16(1): p. 15.

11. Hendricks, K.B. and V.R. Singhal, Does Implementing an Effective TQM Program Actually Improve Operating Performance? Empirical Evidence from Firms That Have Won Quality Awards. Management Science, 1997. 43(9): p. 1258-1274.

12. Hellsten, U. and B. Klefsjo, TQM as a management system consisting of values, techniques and tools. TQM Magazine, 2000. 12(4): p. 238-244.

13. Bunney, H.S. and B.G. Dale, The implementation of quality management tools and techniques: A study. TQM Magazine, 1997. 9(3): p. 183-189.

14. Stephens, B., Implementation of ISO 9000 or Ford's Q1 award: Effects on organizational knowledge and application of TQM principles and quality tools. TQM Magazine, 1997. 9(3): p. 190-200.

15. Boje, D.M. and R.D. Winsor, The resurrection of Taylorism: total quality management's hidden agenda. Journal of Organizational Change Management, 1993. 6(4): p. 57-70.

16. Spector, B. and M. Beer, Beyond TQM programmes. Journal of Organizational Change Management, 1994. 7(2): p. 63-70.

17. Sitkin, S.B., K.M. Sutcliffe, and R.G. Schroeder, Distinguishing Control from Learning in Total Quality Management: A Contingency Perspective. The Academy of Management Review, 1994. 19(3): p. 537-564.

18. Wilkinson, A., et al., Managing With Total Quality Management: Theory and Practice. 1998, London: MacMillan.

19. Zhang, Z., Developing a model of quality management methods and evaluating their effects on business performance. Total Quality Management, 2000. 11(1): p. 129-137.

20. Sousa, R. and C.A. Voss, Quality management re-visited: a reflective review and agenda for future research. Journal of Operations Management, 2002. 20(1): p. 91-109.

21. Zackrisson, J., et al., Quality by a step-by-step program in low scale industries. International Journal of Production Economics, 1995. 41(1-3): p. 419-427.

22. Hakes, J.E., Can measuring results produce results: one manager's view. Evaluation and Program Planning, 2001. 24(3): p. 319-327.

23. Capon, N., M.M. Kaye, and M. Wood, Measuring the success of a TQM programme. The International Journal of Quality & Reliability Management, 1995. 12(8): p. 8.

24. Krishnan, R., et al., In Search of Quality Improvement: Problems of Design and Implementation. The Academy of Management Executive (1993-2005), 1993. 7(4): p. 7-20.

25. Dixon.J.R, Nanni.A.J, and Vollmann.T.E, The new performance chalenge. 1990, Dow-Jones rwin: homewood, IL.

26. GSM. Graduate scool of management. 2011 [cited 2011; Available from: http://gsm.upm.edu.my/gsm-staff-mf.htm.

*****

www.ingramcontent.com/pod-product-compliance
Lightning Source LLC
Chambersburg PA
CBHW080907170526
45158CB00008B/2022